Exploring Combinatorial Mathematics

Richard Grassl & Oscar Levin

1st Edition, Fall 2019

Edition: First Edition

Website: discrete.openmathbooks.org/ecm/

©2018-2019   Richard Grassl & Oscar Levin

Permission is granted to copy, distribute and/or modify this document under the terms of the GNU Free Documentation License, Version 1.3 or any later version published by the Free Software Foundation; with no Invariant Sections, no Front-Cover Texts, and no Back-Cover Texts. A copy of the license is included in the appendix entitled "GNU Free Documentation License".

Acknowledgements

This book is a synthesis of three textbooks on discrete mathematics and combinatorics. First, many of the problems, and indeed entire sections, come directly from *Combinatorics Through Guided Discovery* by Kenneth P. Bogart. This is possible thanks to the Bogart estate's decision to release his book under an open license (the GFDL, the same license this text uses). The format of this book has been made to match that of Bogart's text, with most of the exposition being driven by problems (which we call *activities*) for students to work on. A copy of Bogart's book can be found for free at bogart.openmathbooks.org.

The second source for material is *Discrete and Combinatorial Mathematics* by Richard Grassl and Tabitha Mingus. This is a set of notes used for many years as the textbook for MATH 528 at the University of Northern Colorado. We have tried to capture the flavor and emphasis of this book, and have used many of its exercises among our *activities*.

For the background material contained in the index, as well as much of the chapter on graph theory, material was borrowed from *Discrete Mathematics: an Open Introduction*, the book I wrote for the undergraduate discrete course at the University of Northern Colorado. This book can be found (for free) at discrete.openmathbooks.org.

Finally, thanks to Ricardo Diaz for his helpful suggestions while compiling this book. And thanks in advance to all the students who will soon be contacting me with suggestions for improvements.

Oscar Levin, Ph.D.
University of Northern Colorado, 2019

Preface

This book was written specifically to be used as the textbook for the Master's level Discrete Mathematics course at the University of Northern Colorado. This course is part of a MA in Mathematics with a Teaching Emphasis. Most of the students in the course are current secondary math teachers. This intended audience has influenced the style and content of the book in a few important ways.

First, we acknowledge that not everyone reading this book will be immediately familiar with the content of a standard undergraduate discrete mathematics course. Little is assumed about the reader's previous work in the subject, beyond a general understanding of how abstract mathematics proceeds, as well as some basic ability with mathematical proof. For the reader completely unfamiliar with these and the basic objects of mathematical study (sets and functions), background material is included in an Appendix.

Topics have been selected to illustrate larger concepts of interest to secondary teachers. We have put an emphasis on understanding simple concepts deeply and in more than one way. Although some topics intersect secondary curriculum, most of the questions here are at a higher level. Still, the problem solving strategies and big ideas illustrated by our questions have applications to secondary mathematics. This emphasis is quite different than other mid level discrete and combinatorics textbooks, since we are not preparing our readers to begin a career in research mathematics.

While this course is not a course on teaching mathematics, we have tried to model good pedagogical practice. As you will see, almost all of the textbook consists of *Activities* and *Exercises* that guide students to discover mathematics for themselves. This will require quite a bit more work both from students and instructors, but we strongly believe that the best way to learn mathematics is by doing mathematics. Most of all, discovering mathematics is fun.

How to use this book.

This book is mostly about solving problems, and using those problems and their solutions to develop mathematics and mathematical understanding. The main three chapters consist of more than 300 *activities* and *exercises* which the reader is asked to complete. That is how you should use the book. As you read, when you come to an activity, really take some time to try to solve it. Many of the activities are broken down into tasks to help guide you to a solution. If you get stuck, check for a hint. In the pdf, problems with hints are marked with a small "[hint]" link that will take you to the hint in the back of the book; clicking on the number of the hint will return you to the activity. The online version has expandable hint links under any activities for which they are provided.

The exercises are similar to activities, but can be thought of as optional. In particular, it is less important that you complete the exercise immediately, as they are not used to motivate the next part of the text. Some of the exercises allow you to deepen your understanding of a topic; others are just for practice.

Solutions are not provided explicitly, but often the important content of an activity will be explained further into the book. We have made every effort to make the problems "self checking" so that you do not run the risk of learning something false.

For the most part, if you are unable to complete an activity, assuming you have put a reasonable effort in already, it would be safe to move on. If the activity is meant to guide you to a fundamental concept, that will become apparent, and later activities and exposition will help you uncover that concept. Other activities are meant to provide interesting examples of these big ideas, but missing any one of them will not impede your progress on other activities.

Contents

Acknowledgements		iii
Preface		v

1 Basic Combinatorics 1
  1.1 Pascal's Triangle and Binomial Coefficients 2
  1.2 Proofs in Combinatorics 10
      1.2.1 Two Motivating Examples 10
      1.2.2 Interlude: The Sum and Product Principles 13
      1.2.3 More Proofs 20
      1.2.4 Non-combinatorial Proofs in Combinatorics 25
  1.3 Counting with Equivalence 27
      1.3.1 The Quotient Principle 27
      1.3.2 When does Order Matter? 30
      1.3.3 More Distribution Problems 37
  1.4 Counting with Recursion 40
      1.4.1 Finding Recurrence Relations 40
      1.4.2 Using Recurrence Relations 43
      1.4.3 Exploring the Fibonacci Sequence 48
  1.5 The Catalan Numbers 52
      1.5.1 A Few Counting Problems 52
      1.5.2 The Catalan Numbers 54
      1.5.3 More Problems 56

2 Advanced Combinatorics 59
  2.1 The Principle of Inclusion and Exclusion: the Size of a Union . . . 59
      2.1.1 Unions of two or three sets 59
      2.1.2 Unions of an arbitrary number of sets 60
      2.1.3 The Principle of Inclusion and Exclusion 62
      2.1.4 Application of Inclusion and Exclusion 62
  2.2 Generating Functions 67
      2.2.1 Visualizing Counting with Pictures 68
      2.2.2 Picture functions 69
      2.2.3 Generating functions 70
      2.2.4 Power series 72
      2.2.5 The extended binomial theorem and multisets 74
      2.2.6 Generating Functions and Recurrence Relations 76
      2.2.7 Partial fractions 80
  2.3 Partitions of Sets 86
      2.3.1 Stirling Numbers of the Second Kind 87
      2.3.2 Formulas for Stirling Numbers (of the second kind) 91
      2.3.3 Identities with Stirling Numbers 94
      2.3.4 Bell Numbers 96
  2.4 Partitions of Integers 100
      2.4.1 Partition numbers 100

	2.4.2	Using Geometry	102
	2.4.3	Partitions into distinct parts	107
	2.4.4	Using Generating Functions	108

3 Graph Theory   113
  3.1 Definitions 114
  3.2 Walking Around Graphs 125
       3.2.1 Euler paths and circuits 126
       3.2.2 Proofs for Euler Paths and Circuits 128
       3.2.3 Hamilton Paths 129
  3.3 Planar Graphs and Euler's Formula 131
       3.3.1 Interlude: Mathematical Induction 133
       3.3.2 Proof of Euler's formula for planar graphs. 135
  3.4 Applications of Euler's Formula 137
       3.4.1 Non-planar Graphs 137
       3.4.2 Polyhedra 139
  3.5 Coloring 143
       3.5.1 Coloring in General 146
       3.5.2 Coloring Edges 149
       3.5.3 Ramsey Theory 150
  3.6 Matching in Bipartite Graphs 152

A Group Projects   157
  Project 1: Paths and Binomial Coefficients 158
  Project 2: On Triangular Numbers 159
  Project 3: Fibonacci 160
  Project 4: Generalized Pascal Triangles 161
  Project 5: A Pascal-like Triangle of Eulerian Numbers 162
  Project 6: The Leibniz Harmonic Triangle 163
  Project 7: The Twelve Days of Christmas (in the Discrete Math Style) . 164

B Background Material   165
  B.1 Mathematical Statements 165
       B.1.1 Atomic and Molecular Statements 165
       B.1.2 Implications 167
       B.1.3 Predicates and Quantifiers 174
       B.1.4 Exercises 176
  B.2 Sets 182
       B.2.1 Notation 182
       B.2.2 Relationships Between Sets 185
       B.2.3 Operations On Sets 188
       B.2.4 Venn Diagrams 190
       B.2.5 Exercises 191
  B.3 Functions 195
       B.3.1 Describing Functions 196
       B.3.2 Surjections, Injections, and Bijections 200
       B.3.3 Image and Inverse Image 202
       B.3.4 Exercises 205
  B.4 Propositional Logic 212

		B.4.1	Truth Tables	212
		B.4.2	Logical Equivalence	214
		B.4.3	Deductions	217
		B.4.4	Beyond Propositions	219
		B.4.5	Exercises	221
	B.5	Proofs	225
		B.5.1	Direct Proof	227
		B.5.2	Proof by Contrapositive	228
		B.5.3	Proof by Contradiction	229
		B.5.4	Proof by (counter) Example	231
		B.5.5	Proof by Cases	232
		B.5.6	Exercises	233
	B.6	Induction	237
		B.6.1	Stamps	237
		B.6.2	Formalizing Proofs	239
		B.6.3	Examples	240
		B.6.4	Strong Induction	244
		B.6.5	Exercises	246

C	Hints to selected activities and exercises	253
D	Solutions to selected exercises from the background material	269
E	GNU Free Documentation License	283
F	List of Symbols	289
	Index	291

Chapter 1
Basic Combinatorics

One of the first things you learn in mathematics is how to count. Now we want to count large collections of things quickly and precisely. For example:

- In a group of 10 people, if everyone shakes hands with everyone else exactly once, how many handshakes took place?

- How many ways can you distribute 10 girl scout cookies to 7 boy scouts?

- How many anagrams are there of "anagram"?

Before tackling questions like these, let's look at the basics of counting.

1.1 Pascal's Triangle and Binomial Coefficients

Let's start our investigation of combinatorics by examining Pascal's Triangle. If you are already familiar with this mathematical object, try to see it with fresh eyes. Look at the triangle as if you have no idea where it comes from. What do you notice?

Activity 1

Treating Pascal's Triangle as nothing more than a triangular array of numbers, what do you notice about it? Are there any patterns to the numbers? Try to find some patterns and then see if you can answer (or have already answered) the following:

(a) How are the entries in the triangle found? What would the next row down be?

(b) What is the sum of the entries in the nth row?

(c) What is $1 - 5 + 10 - 10 + 5 - 1$? What pattern is this an example of, and does it always work?

(d) What is $1 + 3 + 6 + 10 + 15$? What pattern is this an example of, and does it always work?

(e) What is $1 + 7 + 15 + 10 + 1$? What pattern is this an example of, and does it always work?

(f) Are there any groups of numbers in the triangle that are particularly recognizable?

We would like to understand why the patterns you discovered above exist, and to prove that they really do exist everywhere in the triangle. There are at least two ways we can proceed (and we will try to proceed in both ways to maximize our understanding). First, we can use notation to describe the entries in the triangle, provide a definition for what each entry is, and prove our results entirely based on these definitions. Second, we can ascribe some *meaning* to the entries in the triangle, saying what the numbers represent, and then make arguments about those representations.

The advantage to pursuing both these approaches is that doing so creates a feedback loop. A fact we establish using pure symbolic manipulation allows us to conclude something about the real world and what these numbers represent. That helps us understand the applications better, which in turn might help us make arguments about those applications, which can then establish a purely symbolic fact about the triangle.

Let's pause and just for fun consider a few discrete math questions that probably have nothing to do with Pascal's Triangle or each other. Probably.

Activity 2

The **integer lattice** is the set of all points in the Cartesian plane for which both the x and y coordinates are integers.

A **lattice path** is one of the shortest possible paths connecting two points on the lattice, moving only horizontally and vertically. For example, here are three possible lattice paths from the points $(0, 0)$ to $(3, 2)$:

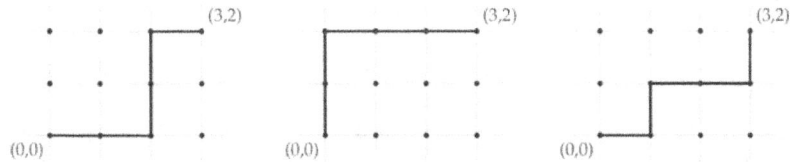

(a) How many lattice paths are there between $(0, 0)$ and $(3, 1)$? Draw or list them all (you might want to invent some notation for describing a path without drawing it).

(b) How many lattice paths are there between $(0, 0)$ and $(2, 2)$? Draw or list them all to be sure.

(c) How many lattice paths are there between $(0, 0)$ and $(3, 2)$? How can you be sure?

Activity 3

Recall that a **subset** A of a set B is a set all of whose elements are also elements of B; we write $A \subseteq B$. For example, some of the subsets of $B = \{1, 2, 3, 4, 5\}$ are $\{2, 4, 5\}$, $\{1, 2, 3, 4, 5\}$ and $\emptyset$. Recall also that the **cardinality** of a set is simply the number of elements in it.

Since we will often consider sets of the form $\{1, 2, 3, \ldots, n\}$, let's adopt the notation $[n]$ for this set.

(a) Write out all subsets of $[3] = \{1, 2, 3\}$. Then group the subsets by cardinality. How many subsets have cardinality 0? How many have cardinality 1? 2? 3?

(b) Write out all subsets of $[4]$, and determine how many have each possible cardinality.

(c) How many subsets of $[5]$ are there of each cardinality? Try to answer this questions without writing out all 32 subsets.

Activity 4

Some definitions: A **bit** is either 0 or 1 (bit is short for "binary digit"). Thus a **bit string** is a string of bits. The **length** of a bit string is the number of bits in the string; the **weight** of a bit string is the number of 1's in the string (or equivalently, the sum of the bits). A n**-bit string** means a bit string of length n.

We will write B_k^n to mean the set of all n-bit strings of weight k. So for example, some of the elements in B_3^5 are,

$$11100 \quad 10101 \quad 01101.$$

(a) Write out all 3-bit strings. Group these by weight. How many strings are there of each weight?

(b) Write out all 4-bit strings. You might want to use the list you had in the previous part as a starting point (how would you do this?). Again, group these strings by weight and see how many there are of each type.

(c) How many 5-bit strings of weight 3 are there? List all of these, and say how they relate to some of the 4-bit strings you found above.

We are not quite ready to consider the algebraic approach yet, but here is a fun algebra exercise.

Activity 5

Multiply out (and collect like terms): $(x + y)^3$. Repeat for $(x + y)^4$ and $(x + y)^5$.

(a) What is the coefficient of $x^3 y^2$ in the expansion of $(x + y)^5$?

(b) What are the coefficients of x^2y^2 and x^3y in the expansion of $(x + y)^4$? How does this relate to the previous question? Why does this make sense?

(c) If you believe the connection between coefficients and Pascal's triangle you have discovered above, find the coefficient of x^7y^4 in the expansion of $(x + y)^{11}$ using Pascal's triangle.

Hopefully by now you have some questions that are begging to be answered. Here are two that we will focus on specifically:

1. Why are the answers to all the counting questions above the same as each other?

2. Why are the answers to all the counting questions above numbers in Pascal's triangle?

We could answer the first question by answering the second: if we can say why each answer is a particular entry in Pascal's triangle, then we know two answers will be equal because they are in the same place on the triangle. But we can do better. In fact, we should be able to explain why these counting questions correspond to each other without knowing *any* of the answers.

Activity 6

(a) Explain why the number of lattice paths from $(0,0)$ to $(k, n - k)$ is the same as number of n-bit strings of weight k. You do not need to give a formal proof here, just explain the idea.

(b) Explain why the number of n-bit strings of weight k is the same as the number of k-element subsets of $[n] = \{1, 2, 3, \ldots, n\}$. Again, an informal argument is fine.

(c) Explain why the number of k-element subsets of $[n]$ is the same as the coefficient of $x^k y^{n-k}$ in the expansion of $(x + y)^n$. [Hint]

Are you satisfied with the explanations you gave above? Do you think they would count as a formal mathematical proof that the counting questions have the same answer? One way to make this sort of an argument more precise is to involve precisely defined mathematical objects.

Activity 7

Consider functions $f : X \to Y$ where X and Y are finite sets. In fact, let's say $X = [4]$. If you need a refresher on basic ideas about functions, check out Section B.3.

(a) Can you say anything at all about Y if you know there is some function $f : X \to Y$? Give some examples of sets Y and functions f.

- **(b)** What if $f : X \to Y$ is *injective* (in other words, one-to-one)? Which of the examples you found above no longer work? What can you conclude about Y?

- **(c)** What if $f : X \to Y$ is *surjective* (i.e., onto)? Now what can you say about Y?

- **(d)** What if $f : X \to Y$ is *bijective* (so both injective and surjective)? State the general principle.

The previous activity was meant to help you discover what is usually called the **bijection principle**: *Two sets X and Y have the same cardinality if and only if there is a bijection $f : X \to Y$.*

If you want to prove that two sets have the same number of elements, all that is required is to define a function from one set to the other, and prove that the function is a bijection. For us, those two sets will be the set of outcomes we are counting. So for example, we can show that the *set* of n-bit strings of weight k has the same cardinality as the *set* of k-element subsets of $[n]$. This will prove that the counting questions, "how many n-bit strings of weight k are there?" and, "how many k-element subsets of $[n]$ are there?" have the same answer.

Example 1.1.1

Define a bijection $f : X \to Y$ where X is the set of lattice paths from $(0,0)$ to $(k, n-k)$ and Y is the set of n-bit strings of weight k. Prove that f really is a bijection.

Solution. This really comes down to how to represent the elements of the set X. We represent a lattice path from $(0,0)$ to $(k, n-k)$ as a sequence of R's and U's, representing a step to the right or a step up. Note that this means that each lattice path will be a sequence containing a total of n symbols, of which k will be R's.

Now we can define the bijection. The function f takes a string of R's and U's, and replaces each R with a 1 and each U with a 0.

To prove this is a bijection, there are four things we need to check:

1. That $f(x)$ is defined for each $x \in X$ (i.e., every lattice path is sent to some n-bit string of weight k).

2. That $f(x)$ is *uniquely* defined for each $x \in X$ (i.e., no lattice path is sent to more than one bit string).

3. That every $y \in Y$ is the image of at least one $x \in X$ (i.e., that every n-bit string of weight k is the output of the function at least once).

4. That every $y \in Y$ is the image of at most one $x \in X$ (i.e., that no n-bit string of weight k is the output of the function more than once.)

The first two conditions guarantee that f will be a function. The third says that f is onto (a surjection). The fourth says that f is one-to-one (an injection).

Let's do this. First, since elements in X contain k R's and $n - k$ U's, applying the rule for f will result in a bit string containing k 1's and $n - k$ 0's, for a total length of n. Thus each element of X will be sent to an element of Y. Further, it is not possible for an element of X to be sent to more than one element of Y since our procedure is well defined.

To show that f is surjective, consider an arbitrary $y \in Y$. If we replace each 1 with an R and each 0 with a U, then we get an element $x \in X$ (because of the length of weight of the strings) and then $f(x) = y$.

Finally, to show that f is injective, consider $x_1, x_2 \in X$ and suppose $x_1 \neq x_2$. Then there is a first position in which they differ (a point on the lattice path where one path goes right and the other goes up). But then this bit will be different in $f(x_1)$ and $f(x_2)$.

Exercise 8
Define a bijection $f : X \to Y$ where X is the set of n-bit strings of weight k and Y is set of k-element subsets of $[n] = \{1, 2, 3, \ldots, n\}$. Again, prove that f is a bijection.

Exercise 9
Can you now conclude that the number of lattice paths from $(0,0)$ to $(k, n - k)$ is the same as the number of k-element subsets of $[n]$? Can you easily give the bijection (and be sure it is really a bijection)? [Hint]

What the example and problems above say is that all these counting questions correspond to each other. But why do the answers always appear in Pascal's triangle? To answer this, we need to start with a definition for the numbers in Pascal's triangle.

First, some notation: let's write $\binom{n}{k}$ for the kth entry in row n. For both n and k, start counting at 0. So for example $\binom{5}{1} = 5$, and $\binom{n}{0} = \binom{n}{n} = 1$ for all n (or at least the ones we can see). We will read $\binom{n}{k}$ as "n choose k" for reasons that will become clear soon.

Now we need to decide what $\binom{n}{k}$ should be defined as. It is not enough to simply define it as the specific entry in Pascal's triangle, since we haven't yet defined Pascal's triangle (we will want to define Pascal's triangle as the triangle of the numbers $\binom{n}{k}$). Here are some choices:

Possible Definitions for $\binom{n}{k}$.

For each integer $n \geq 0$ and integer k with $0 \leq k \leq n$ the number

$$\binom{n}{k},$$

read "n choose k" is:

1. the number of n-bit strings of weight k, that is, $\binom{n}{k} = |B_k^n|$.

2. $\binom{n}{k}$ is the number of subsets of a set of size n each with cardinality k.

> 3. $\binom{n}{k}$ is the number of lattice paths of length n containing k steps to the right.
>
> 4. $\binom{n}{k}$ is the coefficient of $x^k y^{n-k}$ in the expansion of $(x+y)^n$.
>
> 5. $\binom{n}{k}$ is the number of ways to select k objects from a total of n objects.
>
> 6. $\binom{n}{0} = \binom{n}{n} = 1$ and for all $n \geq 1$ and $1 < k < n$ we have $\binom{n}{k} = \binom{n-1}{k-1} + \binom{n-1}{k}$.

You have already proved that choices 1, 2 and 3 are equivalent, and we have good reason to believe these are also equivalent to choice 4.

What about choice 5? This is new, although not really, as it captures all of the previous four: Each of our counting problems above can be viewed in this way:

- How many bit strings have length 5 and weight 3? We must choose 3 of the 5 bits to be 1's. There are $\binom{5}{3}$ ways to do this, so there are $\binom{5}{3}$ such bit strings.

- How many subsets of $\{1, 2, 3, 4, 5\}$ contain exactly 3 elements? We must choose 3 of the 5 elements to be in our subset. There are $\binom{5}{3}$ ways to do this, so there are $\binom{5}{3}$ such subsets.

- How many lattice paths are there from (0,0) to (3,2)? We must choose 3 of the 5 steps to be towards the right. There are $\binom{5}{3}$ ways to do this, so there are $\binom{5}{3}$ such lattice paths.

- What is the coefficient of $x^3 y^2$ in the expansion of $(x+y)^5$? We must choose 3 of the 5 copies of the binomial to contribute an x. There are $\binom{5}{3}$ ways to do this, so the coefficient is $\binom{5}{3}$.

In fact, choice 5 is usually taken as the definition of $\binom{n}{k}$, which is why we say *n choose k*. Although interestingly enough, the other common name for $\binom{n}{k}$ is a **binomial coefficient**, which refers to the coefficients of the binomial $(x+y)^n$.

What does all this have to do with Pascal's triangle though? You might think of the entries in the triangle as the result of adding the two entries above it. That is, Pascal's triangle contains those numbers described by choice 6 above.

> **Recurrence relation for $\binom{n}{k}$.**
>
> For any $n \geq 1$ and $0 < k < n$:
>
> $$\binom{n}{k} = \binom{n-1}{k-1} + \binom{n-1}{k}.$$

The last piece of the puzzle is to connect this definition to all the others. Of course given the work done above, it would be enough to prove that any one of the models for binomial coefficients match the recurrence for Pascal's triangle, but it is instructive to check all four.

Example 1.1.2

Prove that bit strings satisfy the same recurrence as Pascal's triangle. That is prove that $|B^n_k| = |B^{n-1}_{k-1}| + |B^{n-1}_k|$.

Solution. Partition the n-bit strings of weight k into two disjoint sets: those that start with a 0 and those that start with a 1. Consider the remainder of each bit string (after that first bit). Each n-bit string that starts with a 1 ends with an $(n-1)$-bit string of weight $k-1$. There are $|B^{n-1}_{k-1}|$ of these. Each n-bit string that starts with a 0 ends with an $(n-1)$-bit string of weight k. There are $|B^{n-1}_k|$ of these. So all together $|B^n_k| = |B^{n-1}_{k-1}| + |B^{n-1}_k|$.

Exercise 10

(a) Prove that subsets satisfy the same recurrence as Pascal's triangle. Make sure you clearly state what you are proving.

(b) Prove that lattice paths satisfy the same recurrence as Pascal's triangle. What exactly do you need to prove here?

(c) Prove that binomial coefficients (the actual coefficients of the expansion of the binomial $(x+y)^n$) satisfy the same recurrence as Pascal's triangle.

At last we can rest easy that we can use Pascal's triangle to calculate binomial coefficients and as such find numeric values for the answers to counting questions. How many pizza's containing three distinct toppings can you make if the pizza place offers 10 toppings? Well the answer should be $\binom{10}{3}$, and Pascal's triangle says that is 120.

1.2 Proofs in Combinatorics

In this section, we will consider a few proof techniques particular to combinatorics. Along the way we will establish some basic counting principles and further develop our understanding of binomial coefficients.

Why do we care about proofs? It is not because we doubt the truth of so called theorems or wish to establish new results to further mathematics as an academic discipline (although there is nothing wrong with doing that as well). Rather, it is through creating proofs that we gain a deeper understanding of the big ideas present in this subject.

1.2.1 Two Motivating Examples

To get started, let's consider two typical statements in combinatorics which we might wish to prove.

> **Theorem 1.2.1**
> For all $n \geq 0$ and all $0 \leq k \leq n$ we have $\binom{n}{k} = \binom{n}{n-k}$. In other words, Pascal's triangle is symmetric reflected over its vertical altitude.

> **Theorem 1.2.2**
> For all $n \geq 0$, we have $\sum_{k=0}^{n} \binom{n}{k} = 2^n$. That is, the sum of the entries in the nth row of Pascal's triangle is 2^n.

These are just two of the beautiful identities hidden in Pascal's triangle. But why are they true?

There are at least four very distinct proofs we could give for each of these theorems. The next activities and examples will walk you through a few.

Note: throughout this section especially, hints to problems are available. Try the problem first, and if you get stuck, peek at the hint.

First, let's try some proofs of Theorem 1.2.1

> **Activity 11**
> First, an algebraic proof. We will see later that $\binom{n}{k} = \frac{n!}{k!(n-k)!}$, where $r! = r \cdot (r-1) \cdot (r-2) \cdots 2 \cdot 1$ (read "r factorial"). Using this algebraic formula, prove the identity $\binom{n}{k} = \binom{n}{n-k}$. [Hint]

Note that this proof was not by any means difficult, but it did rely on an algebraic identity we have yet to prove. We will see in Activity 14 that algebraic proofs can be very messy. What's more, algebraic proofs rarely explain *why* an identity is true. They lack meaning.

Slightly more meaning can be found in proofs by induction. There at least you see the connection to the recursion defining the terms. We give an example of such an argument here.

Example 1.2.3

We claim the identity $\binom{n}{k} = \binom{n}{n-k}$ is true for all $n \geq 0$, so induction would be a natural proof technique to try. Give a proof by mathematical induction on n.

Solution. We will do induction on n. For $n = 0$, the only value of k that is interesting is $k = 0$, and here $\binom{n}{k} = \binom{0}{0} = \binom{n}{n-k}$. Thus the base case is established.

For the inductive case, fix an arbitrary n and assume that for all k with $0 \leq k \leq n-1$ we have $\binom{n-1}{k} = \binom{n-1}{(n-1)-k}$. Consider $\binom{n}{k}$ for an arbitrary $0 \leq k < n$. By the recurssion for Pascal's triangle, we have

$$\binom{n}{k} = \binom{n-1}{k-1} + \binom{n-1}{k}.$$

Now apply the induction hypothesis to the right hand side. We get

$$\binom{n-1}{k-1} + \binom{n-1}{k} = \binom{n-1}{n-1-(k-1)} + \binom{n-1}{n-1-k} = \binom{n-1}{n-k} + \binom{n-1}{n-k-1}.$$

Then using the Pascal recurrence again, we get

$$\binom{n-1}{n-k} + \binom{n-1}{n-k-1} = \binom{n}{n-k}.$$

Thus $\binom{n}{k} = \binom{n}{n-k}$, as long as $0 \leq k < n$. The case where $k = n$ is trivial as $\binom{n}{n} = \binom{n}{0} = 1$.

The next two proofs are examples of **combinatorial proofs**.

Activity 12
Think about what $\binom{n}{k}$ counts.

(a) Suppose you own n bow ties, and want to choose k of them to take on a trip. How many different collections of bow ties can you take? [Hint]

(b) Why is the answer to the above question also $\binom{n}{n-k}$? [Hint]

(c) Why are the two parts above enough to establish the identity $\binom{n}{k} = \binom{n}{n-k}$?

Finally, here is a subtly different proof.

Activity 13
Consider the set $[n] = \{1, 2, \ldots, n\}$ and its power set $\mathcal{P}([n])$ of all subsets of $[n]$. Define the function $f : \mathcal{P}([n]) \to \mathcal{P}([n])$ by $f(A) = [n] \setminus A$ for any $A \in \mathcal{P}([n])$ (that is, $f(A)$ is the complement of A in $[n]$).

(a) Prove that f is a bijection. In fact, f is an **involution** in that it is its own inverse.

(b) For any set $A \in \mathcal{P}([n])$, if $|A| = k$, what is the $|f(A)|$?

(c) Is f still a bijection if we restrict it to just the subsets of $[n]$ that have size k? Perhaps we should also restrict our codomain (to what?). Explain.

(d) Why is the above enough to establish the identity $\binom{n}{k} = \binom{n}{n-k}$? [Hint]

How are Activity 12 and Activity 13 different? The first consists of answering a single counting question in two different ways, and those ways are the two sides of the identity. Such proofs are sometimes called *double counting proofs*, or sometimes just *combinatorial proofs*. On the other hand, Activity 13 proceeds by showing two different sets have the same size, using a bijection, and shows that these two sets are counted by each side of the identity. These proofs are called *bijective proofs* (and are also sometimes grouped together with double counting proofs as combinatorial proofs).

Here is yet another combinatorial proof of the identity $\binom{n}{k} = \binom{n}{n-k}$. We give both double counting and bijective variants.

Example 1.2.4
Prove that $\binom{n}{k} = \binom{n}{n-k}$ using bit strings.

Solution. $\binom{n}{k}$ counts the number of n-bit strings of weight k. This is because of the n bits, we must choose k to be 1's, which can be done in $\binom{n}{k}$ ways. The number of n-bit strings of weight k is also $\binom{n}{n-k}$, since of the n bits, we must choose $n - k$ to be 0's (making the remaining k 1's), which can be done in $\binom{n}{n-k}$ ways. Thus $\binom{n}{k} = \binom{n}{n-k}$.

The above is a complete proof, but really $\binom{n}{k}$ can be thought of as *defined* as the size of the set B_k^n. But then $\binom{n}{n-k}$ is the size of the set B_{n-k}^n. So what we are really saying here is that $|B_k^n| = |B_{n-k}^n|$, which is true because the function that switches 0's to 1's and 1's to 0's in a bit string is a bijection between these two sets.

Both styles of combinatorial proof have the advantage that they do an excellent job of illustrating what is really going on in an identity. They reinforce our understanding of how to solve counting problems and our understanding of basic combinatorial principles.

This is not to say that other proof techniques, especially mathematical induction, are not also helpful. Inductive proofs demonstrate the importance of the recursive nature of combinatorics. Even if we didn't know what Pascal's triangle told us about the real world, we would see that the identity was true entirely based on the recursive definition of its entries.

Now here are four proofs of Theorem 1.2.2.

Activity 14
Using the formula $\binom{n}{k} = \frac{n!}{k!(n-k!)}$, you should be able to find a common denominator in the sum $\sum_{k=0}^{n} \binom{n}{k}$ and show that this simplifies to 2^n. [Hint]

Activity 15
We wish to establish the identity $\sum_{k=0}^{n} \binom{n}{k} = 2^n$ for all natural numbers n, so it would be natural to give a proof by induction. Do this. [Hint]

There are lots of ways to give a combinatorial proof. Here is one.

Activity 16
Consider the question: how many subsets of $[n]$ are there?

(a) How many subsets have cardinality 0? How many have cardinality 1? And so on. How many would all of these give us all together? [Hint]

(b) Explain why the answer is also 2^n. [Hint]

Conclude: Since each side of the identity is the answer to the same counting question, we have established the identity.

We will not give a bijective proof, although you might include this as part of the above if you didn't see why 2^n was the number of subsets of $[n]$. Instead, you might see directly that 2^n is the number of n-bit strings (by writing the numbers 0 through $2^n - 1$ in binary) and then define a bijection from n-bit strings to subsets of $[n]$.

Finally, here is a clever proof of the identity.

Activity 17
What does the binomial theorem say again? That is, expand $(x + y)^n$. What happens when you substitute $x = y = 1$? [Hint]

This use of the binomial theorem is an example of one of the many uses for *generating functions* which we will return to later. For now, you might enjoy plugging in other values to the binomial theorem to uncover new binomial identities.

1.2.2 Interlude: The Sum and Product Principles

Any proof you write in mathematics must assume some foundational principles. In Euclidean geometry, you have axioms, common notions, and often other propositions, used as justification in your proof. Let's pause to consider two foundational principles in combinatorics that, while intuitively obvious, are worth pointing out as foundational.

Activity 18

(a) A restaurant offers 8 appetizers and 14 entrées. How many choices do you have if:

  (a) you will eat one dish, either an appetizer or an entrée?

  (b) you are extra hungry and want to eat both an appetizer and an entrée?

(b) Think about the methods you used to solve question (a). Write down the rules for these methods.

(c) Do your rules work? A standard deck of playing cards has 26 red cards and 12 face cards.

  (a) How many ways can you select a card which is either red or a face card?

  (b) How many ways can you select a card which is both red and a face card?

  (c) How many ways can you select two cards so that the first one is red and the second one is a face card?

When solving a counting question, it is often advantageous to break the problem into parts. If you think of the counting question as asking how many ways there are to complete a task, then you might break that task up in various ways, count the number of ways to complete each sub-task, and them combine those somehow to get the number of ways to complete the overall task. How do you combine them?

> **Sum Principle.**
>
> The **sum principle** states that if task A can be completed in m ways, and task B can be completed in n *disjoint* ways, then completing "A or B" is a task that can be completed in $m + n$ ways.

> **Product Principle.**
>
> The **product principle** states that if task A can be completed in m ways, and each possibility for A allows for exactly n ways for task B to be completed, then the task "A and B" (or "A followed by B") can be completed in $m \cdot n$ ways.

Note, both the sum and product principles are true for more than two sub-tasks. You could prove this easily by induction using the principles for two sets as your base case. But how do we know the principles for two sets are true?

To make the sum and product principles more mathematically rigorous, consider operations on sets.

Proposition 1.2.5
If A and B are finite, disjoint sets (so $A \cap B = \emptyset$) then $|A \cup B| = |A| + |B|$.

This suggests that the union of (disjoint) sets somehow corresponds to the *disjunction* (or) of completing tasks. You might guess that the set operation that corresponds to *conjunction* (and) of tasks is set intersection. But this cannot be correct: the intersection of two sets is no larger than the smallest of the two sets, while combining tasks with conjunction apparently corresponds to multiplication.

A better way to think of the conjunction is as *concatenation*. You are performing one task, *followed* by the other. Recall that $A \times B = \{(a,b) : a \in A, b \in B\}$ is the **Cartesian product** of A and B; the set of all ordered pairs where the first element comes from A and the second from B. This is exactly what we need.

Proposition 1.2.6
If A and B are finite sets, then $|A \times B| = |A| \cdot |B|$

We could prove these two propositions without too much trouble. Then we must explain what they have to do with the sum and product principles.

Activity 19

(a) Explain why Proposition 1.2.5 is true.

(b) An incomplete deck of cards only has four face cards (the Jack, King and Queen of diamonds and the King of hearts) and five black cards (the 2 through 6 of spades). A magician asks you to pick a card. How many choices do you have?

(c) How exactly does the counting question above relate to Proposition 1.2.5? How exactly does it relate to the Sum Principle? Illustrate both by listing specific outcomes.

(d) Make explicit the connection between the Sum Principle and Proposition 1.2.5. That is, justify the Sum Principle in terms of cardinality of sets.

Activity 20

(a) Explain why Proposition 1.2.6 is true. Your explanation should make use of Proposition 1.2.5.

(b) Using the same incomplete deck of cards (containing the Jack, King and Queen of diamonds, King of hearts, and 2 through 6 of spades), the magician asks you to put any face card face down under any face down black card. How many choices do you have for this two card stack?

(c) How exactly does the counting question above relate to Proposition 1.2.6? How exactly does it relate to the Product Principle? Illustrate both by listing specific outcomes.

(d) Make explicit the connection between the Product Principle and Proposition 1.2.6. That is, justify the Product Principle in terms of cardinality of sets.

The previous activity might obscure an important subtlety. Perhaps you said that the set A corresponds to the face cards (or to the set of outcomes of the task of selecting a face card) and the set B corresponded to the black cards. The set of outcomes of picking a two card stack is exactly then $A \times B$, so there are $4 \cdot 5 = 20$ outcomes. This is completely correct. But does it generalize?

Activity 21
Suppose the magician hands you 4 cards (the Ace, King, Queen, and Jack of clubs) and asks you to select any two cards (placing them face down in a stack). How many choices do you have? Assume the order the cards are placed down matters.

(a) If you divide the task into two sub-tasks, what are they? How many ways can you complete the first task? How many ways can you complete the second? What is the final answer?

(b) If you think of this problem as counting a Cartesian product, what is set A? What is set B? What is wrong?

The Product Principle is not exactly analogous to finding the cardinality of a Cartesian product after all, since tasks A and B do not need to be disjoint. How do we make sense of the Product Principle then?

Activity 22

(a) Write out all stacks of cards that are possible outcomes in Activity 21. Arrange these into a rectangle of appropriate dimensions.

(b) How might you group the outcomes in a meaningful way? How many groups and how many outcomes in each group?

(c) Find some sets A and B and a bijection between the outcomes (stacks of 2 cards) and the set $A \times B$. Describe how this would work in general.

To illustrate just how fundamental these two principles are, here are a few counting problems that can be solved with them. For each of these, say exactly where and how the sum and product principles are used.

Exercise 23
We have seen that the number of lattice paths from $(0,0)$ to (x,y) is given by $\binom{x+y}{x}$. Now with the sum and product principles, we can answer lots of questions about lattice paths.

(a) How many lattice paths from $(0,0)$ to $(10,10)$ pass through the point $(4,7)$? [Hint]

(b) How many lattice paths from $(0,0)$ to $(10,10)$ do NOT pass through the point $(4,7)$? In what way does this problem use the sum principle? [Hint]

(c) How many lattice paths from $(0,0)$ to $(10,10)$ pass through $(4,7)$ or $(7,4)$?

(d) Explain why can you NOT use the sum principle to answer this question: How many lattice paths from $(0,0)$ to $(10,10)$ pass through $(4,7)$ or $(7,8)$? What is the answer? [Hint]

The last part of the previous problem illustrates what can go wrong if the sets A and B are not disjoint. In general, we have,

$$|A \cup B| = |A| + |B| - |A \cap B|.$$

This is an example of the *Principle of Inclusion/Exclusion*, which we will investigate further later.

This next set of questions you can answer entirely using the product principle. While all the activities ask questions with a general n and k, it would be a good idea to first try to answer the corresponding question with specific numbers and then generalize.

Exercise 24
How many subsets does a set S with n elements have? [Hint]

Exercise 25
A tennis club has $2n$ members. We want to pair up the members by twos for singles matches.

(a) In how many ways may we pair up all the members of the club? (Hint: consider the cases of 2, 4, and 6 members.) [Hint]

(b) Suppose that in addition to specifying who plays whom, for each pairing we say who serves first. Now in how many ways may we specify our pairs?

Exercise 26

A roller coaster car has n rows of seats, each of which has room for two people. If n men and n women get into the car with a man and a woman in each row, in how many ways may they choose their seats? [Hint]

Example 1.2.7
Here are two counting problems, easier than the last two activities, that will lead us to an important general principle.

1. In how many ways can we pass out k distinct pieces of fruit to n children (with no restriction on how many pieces of fruit a child may get)?

2. Assuming $k \leq n$, in how many ways can we pass out k distinct pieces of fruit to n children if each child may get at most one? What is the number if $k > n$? Assume for both questions that we pass out all the fruit.

Solution.

1. We must decide on a child for each piece of fruit. There are n choices for the first piece of fruit, n choices for the second, and so on. The product principle says that there will be n^k ways to distribute the fruit.

    Slightly more mathematically speaking, we are counting the number of functions from a set of cardinality k (the fuits) to a set of cardinality n (the children). Note that it does not make sense to think of the ways we can send the children to the fruit, since we do not require that every child gets a fruit, and we insist that all fruit is distributed.

2. The only thing that changes for the second part is that no child can get more than one piece of fruit. So while we have n choices for the first piece of fruit, we only have $n - 1$ for the second, and $n - 2$ for the third. Continuing until all the fruit is gone gives us a total of,

    $$n \cdot (n - 1) \cdot (n - 2) \cdots (n - k + 1) = \frac{n!}{(n - k)!}$$

    ways to distribute the fruit.

    If $k > n$, then the left-hand-side still works: one of the terms will be 0, and indeed, if there is more fruit than children, there is no way to pass out all the fruit while ensuring that each kid gets at most one fruit.

The last question is an example of a **permutation**. We are asking how to arrange k of the n kids (arranged by which fruit they get). There are $n!$ ways to arrange n elements, so there are $n!$ permutations of n elements. Often we only want to arrange some of these, so we would consider a k-permutation. We will use this often enough that it deserves its own notation.

> **k-permutations of n elements.**
>
> $P(n,k)$ is the number of k-permutations of n elements, the number of ways to *arrange* k objects chosen from n distinct objects.
>
> $$P(n,k) = n \cdot (n-1) \cdot (n-2) \cdots (n-k+1) = \frac{n!}{(n-k)!}.$$

Some other examples of questions where the answer is a permutation:

- How many 5-letter "words" (i.e., strings of symbols) contain no repeated letters? There are 26 choices for the first letter, 25 for the second, 24 for the third, 23 for the fourth and 22 for the fifth, so there are $26 \cdot 25 \cdot 24 \cdot 23 \cdot 22 = P(26,5)$ such words.

- How many different shuffled decks of cards are possible? There are 52 choices for the bottom card, 51 for the next, and so on. So the answer is $P(52,52) = 52!$

- How many *injective* functions from $[k]$ to $[n]$ are there? There are n choices for where 1 is sent, $n-1$ choices for where 2 is sent (since the function is injective, 2 cannot be sent to the same element as 1), and so on. So the answer is $P(n,k)$.

While we are looking at permutations, let's also consider their counterpart, *combinations*. The key distinction is that while a permutation counts different arrangements as different outcomes, a combination does not. Combinations are *selecting*, while permutations are *arranging* (and also selecting).

> **Activity 27**
>
> Suppose you have 17 books strewn around your room. You resolve to get organized.
>
> (a) You decide to throw 10 books in the trash. How many choices do you have for which books to get rid of? [Hint]
>
> (b) You think twice about tossing your books, and you find a bookshelf that can hold 10 books. How many ways can you fill the bookshelf?
>
> (c) How many ways could you fill the 10 book bookshelf if the books were arranged alphabetically? [Hint]

Hopefully you realized that the number of k-combinations of n elements is just $\binom{n}{k}$. This makes sense: you just need to choose which elements to take, you don't need to arrange them.

Notice though that combinations and permutations *are* related in some way. What is that way? Well, once you have chosen k elements from an n-element set, you could then *arrange* those k elements to get a permutation.

Activity 28

How many 5-letter words can be made with distinct letters?

(a) Explain why the answer is $P(26,5)$.

(b) Explain why the answer is also $\binom{26}{5} \cdot 5!$.

(c) Express $P(26,5)$ as the quotient of two factorials. Use this to find a numerical value for $\binom{26}{5}$. [Hint]

There is nothing special about 26 and 5 in the above activity.

Theorem 1.2.8

$P(n,k) = \binom{n}{k} \cdot k!$, and therefore

$$\binom{n}{k} = \frac{n!}{(n-k)! \cdot k!}.$$

Exercise 29

Give a combinatorial proof for the identity $P(n,k) = \binom{n}{k} \cdot k!$, thus proving Theorem 1.2.8.

Notice that the only thing we needed to find the algebraic formula for binomial coefficients was the product principle and a willingness to solve a counting problem in two ways. In Section 1.3 we will consider this formula again from the other direction.

There are non-combinatorial proofs of this formula as well.

Exercise 30

Use the binomial theorem and the fact that $(x+1)^n = (x+1)^{n-1} + x(x+1)^{n-1}$ to derive the factorial formula

$$\binom{n}{k} = \frac{n!}{(n-k)!k!}.$$

1.2.3 More Proofs

We will use the proof techniques of double counting and bijections throughout the rest of the book, but for now, let's practice a bit. The activities below ask you to give a combinatorial proof, and often also an alternative proof that has some nice features.

Example 1.2.9

Prove the identity $\binom{2n}{2} = 2\binom{n}{2} + n^2$.

Hint. Think of selecting two balls from a set of n distinct red balls and n distinct green balls.

Solution. Split the $2n$ objects into two groups A and B each of size n.

First, you can choose 2 objects from a set of $2n$ objects in $\binom{2n}{2}$ ways. Alternatively, you could select two from group A in $\binom{n}{2}$ ways or two from group B in $\binom{n}{2}$ ways or take one from each in $n \cdot n = n^2$ ways. Now add $\binom{n}{2} + \binom{n}{2} + n \cdot n$ and the result follows.

The reader should attempt an algebraic proof of the above result using the factorial formula for $\binom{n}{k}$.

Exercise 31

Prove the identity $\binom{m+n}{2} - \binom{m}{2} - \binom{n}{2} = mn$. [Hint]

You could also attempt an algebraic proof or perhaps a geometric proof making use of figures consisting of triangular numbers (we will return to this identity at the end of the section to illustrate this). Also, as a challenge, the reader could formulate a similar result involving $\binom{a+b+c}{3}$ and a corresponding proof.

Exercise 32

We have already proved this identity: $\binom{n}{0} + \binom{n}{1} + \binom{n}{2} + \ldots + \binom{n}{n} = 2^n$. Now try giving a combinatorial proof that uses lattice paths. [Hint]

Becoming an expert in combinatorial proofs will help you solve actual counting problems. For example, you might wonder how many positive integers have their digits in strictly increasing order.

Exercise 33

Prove that the number of positive integers that have their digits in strictly increasing order is $2^9 - 1$. Include single digit numbers. [Hint]

Now back to some combinatorial proofs.

Example 1.2.10

Prove $\binom{n}{1} + 2\binom{n}{2} + 3\binom{n}{3} + \ldots + n\binom{n}{n} = n2^{n-1}$.

Solution. Given a set of n people we can select a committee of size k along with a chair from that committee in $k\binom{n}{k}$ ways. We can select a committee (of size 1, or size 2, or ...) and its chair in $\binom{n}{1} + 2\binom{n}{2} + 3\binom{n}{3} + \ldots + n\binom{n}{n}$ ways. Alternatively, we can explain the term $n2^{n-1}$ as follows: choose one of the n people to chair any of the 2^{n-1} subsets of the remaining $n-1$ people.

Some other ways to approach the previous identity might be one or more of the following: $n2^{n-1}$ looks like a derivative, so try differentiating $(1+x)^n$; a reverse and add approach also works; or, first prove $k\binom{n}{k} = n\binom{n-1}{k-1}$ and then use it.

> **Exercise 34**
> Prove the Pascal recurrence: $\binom{n}{k} = \binom{n-1}{k-1} + \binom{n-1}{k}$. Do this in two ways, first with a double counting style proof, and then a bijective proof. [Hint]

The above activity illustrates another important reason to use combinatorial proofs: to establish recurrence relations. The next identity is also a recurrence, this time about *derangements*.

> **Example 1.2.11**
> A **derangement** of a set of elements is a permutation of those elements so that no element appears in its original position. For example, a derangement of $[4]$ might be $3, 1, 4, 2$, but the permutation $3, 2, 4, 1$ is not a derangement because 2 is still in the second position.
> Let d_n denote the number of derangements of $[n]$. We will take $d_0 = 1$ and $d_1 = 0$. Prove that $d_n = (n-1)(d_{n-1} + d_{n-2})$ for $n \geq 2$.
>
> **Solution.** In forming a derangement of $1, 2, 3, \ldots, n$ the integer n can be placed in any of the $n-1$ spots $1, 2, 3, \ldots, n-1$, say spot i. If i goes into spot n there are d_{n-2} ways to finish it. If i does not go into spot n there are d_{n-1} ways to complete the derangement.

Later we will use the Principle of Inclusion/Exclusion to derive a formula for d_n from which the new recursion $d_n = nd_{n-1} + (-1)^n$, and the above recursion, can be derived.

Let's get back to binomial identities.

> **Example 1.2.12**
> Prove, $\binom{n}{0}^2 + \binom{n}{1}^2 + \binom{n}{2}^2 + \ldots + \binom{n}{n}^2 = \binom{2n}{n}$.
>
> **Hint.** First note that $\binom{n}{k}^2 = \binom{n}{k}\binom{n}{n-k}$, by the symmetry of Pascal's triangle. Then, there is a nice proof using lattice paths here, or you select n things from a collection that has n things of one type and n of another.
>
> **Solution.** Given a group of $2n$ people consisting of n men and n women, in how many ways can one choose a group of n people? The answer to that question is just $\binom{2n}{n}$, the right side of the identity in question. One could also form the group of n people in the following way: choose 0 men and n women in $\binom{n}{0}\binom{n}{n} = \binom{n}{0}^2$ ways, or, choose 1 man and $n-1$ women in $\binom{n}{1}\binom{n}{n-1} = \binom{n}{1}^2$ ways, or choose 2 men and $n-2$ women in $\binom{n}{2}\binom{n}{n-2} = \binom{n}{2}^2$ ways and so on. Now add these disjoint cases.
>
> An alternate algebraic proof is less interesting: Extract the coefficient of x^n from both sides of $\left[(x+1)^n\right]^2 = (x+1)^{2n}$.

Here is a interesting use of the previous identity, inside an unexpected proof.

Example 1.2.13
Find a nice expression for the sum
$$\binom{n}{1}^2 + 2\binom{n}{2}^2 + 3\binom{n}{3}^2 + \cdots + n\binom{n}{n}^2.$$

Solution. We use the "reverse and add technique", which we will see in Subsection 1.4.2 is usually reserved for summing arithmetic sums. Surprisingly, it works here.

$$\begin{aligned} S = & \quad \binom{n}{1}^2 + 2\binom{n}{2}^2 + 3\binom{n}{3}^2 + \cdots + (n-1)\binom{n}{n-1}^2 + n\binom{n}{n}^2 \\ + \; S = & \; n\binom{n}{n}^2 + (n-1)\binom{n}{n-1}^2 + \cdots + 2\binom{n}{2}^2 + \binom{n}{1}^2 \\ \hline 2S = & \; n\binom{n}{0}^2 + n\binom{n}{1}^2 + n\binom{n}{2}^2 + n\binom{n}{3}^2 + \cdots + n\binom{n}{n-1}^2 + n\binom{n}{n}^2 \end{aligned}$$

This uses the identity $\binom{n}{k} = \binom{n}{n-k}$. Now factor out an n to get

$$2S = n\left[\binom{n}{0}^2 + \binom{n}{1}^2 + \binom{n}{2}^2 + \cdots + \binom{n}{n}^2\right].$$

Apply the identity in Example 1.2.12 and divide by 2 to get

$$S = \frac{n}{2}\binom{2n}{n}.$$

The next identity is an example of the famed **Hockey Stick Theorem** (find this identity in Pascal's triangle!).

Exercise 35
Prove $\binom{2}{2} + \binom{3}{2} + \binom{4}{2} + \ldots + \binom{n}{2} = \binom{n+1}{3}$. [Hint]

For a non-combinatorial proof of the Hockey Stick Theorem, a proof by mathematical induction is an easy option. An algebraic approach is not!

If you spend enough time working on these proofs, you might find yourself discovering new identities that use the same basic setup. You can take the proof you gave for the Hockey Stick Theorem and with a small tweak, get the following.

Example 1.2.14
Prove $1 \cdot n + 2 \cdot (n-1) + 3 \cdot (n-2) + \cdots + n \cdot 1 = \binom{n+2}{3}$

Solution. You should try to use bit strings again, but this time think about where the *second* 1 (from the left) in the bit string could go. Or for variety, why not try asking about subsets of $[n+2]$?

Let $S = \{1, 2, \ldots, n+2\}$. The number of subsets of S of size 3 is $\binom{n+2}{3}$. Each one looks like $\{a, b, c\}$ with $a < b < c$. Let's count these by looking at the size of the middle element b. If $b = 2$, there is one choice for a, namely $a = 1$ and n choices for c for a total of $1 \cdot n$. If $b = 3$ there are 2 choices for a, and $n - 1$ choices for c for a total of $2(n-1)$. If $b = 4$ the total is $3(n-2)$, and so on. The total derived by looking at cases is

$1 \cdot n + 2(n-1) + 3(n-2) + \ldots + n \cdot 1$ and this must equal $\binom{n+2}{3}$ since the cases are disjoint.

Exercise 36
The proof using subsets we gave for the previous identity asked about 3-element subsets of $[n+2]$. What if we looked at 5-element subsets of $[n+3]$ and considered which the middle (third) number was? Write a binomial identity, with a proof, that uses this idea. [Hint]

Exercise 37
$1 \cdot 1! + 2 \cdot 2! + 3 \cdot 3! + \ldots + n \cdot n! = (n+1)! - 1$. [Hint]

The previous identity can also be established using an inductive proof, or through a clever collapsing sum as follows.

Example 1.2.15
To compute the sum $1 \cdot 1! + 2 \cdot 2! + \cdots + n \cdot n!$, note that $(n+1)! - n! = n \cdot n!$ (just factor out an $n!$). This then makes the sum

$$(2! - 1!) + (3! - 2!) + \cdots + ((n+1)! - n!) = (n+1)! - 1.$$

Exercise 38
$k\binom{n}{k} = n\binom{n-1}{k-1}$. [Hint]

Exercise 39
$\binom{n}{0}d_0 + \binom{n}{1}d_1 + \binom{n}{2}d_2 + \ldots + \binom{n}{n}d_n = n!$ where d_n denotes the nth derangement number, $d_0 = 1, d_1 = 0$. [Hint]

To close this section, let's return to a binomial identity we considered before, and consider a few additional proofs.

Example 1.2.16
We give three and a half proofs of the identity $\binom{m+n}{2} - \binom{m}{2} - \binom{n}{2} = mn$.

1. How many lines that are not vertical or horizontal can you form by connecting the points that are on the positive x and y axes?

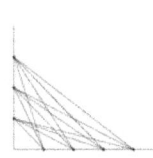
Since we really want only lines that are formed by connecting the m points with the n points, we could say we have mn lines. Or we could consider all $m + n$ points and pick 2 in $\binom{m+n}{2}$ ways and then delete those choices where you took 2 from m or 2 from n, since these formed vertical and horizontal lines, respectively. Thus $\binom{m+n}{2} - \binom{m}{2} - \binom{n}{2} = mn$.

2. There are mn one by one squares in the subdivided m by n rectangle.

    Each choice of arrows (one horizontal, one vertical) specifies one of these squares. Pick two arrows but don't take two from the top or two from the side. Then $\binom{m+n}{2} - \binom{m}{2} - \binom{n}{2} = mn$.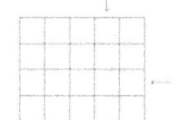

3. Take m people in one group and n in another. How many handshakes can be accomplished? Among the m people there are $\binom{m}{2}$ handshakes; among the n people there are $\binom{n}{2}$ handshakes and between the two groups, mn. But, $\binom{m+n}{2}$ also represents the total number of handshakes among the people. Then we get: $\binom{m}{2} + \binom{n}{2} + mn = \binom{m+n}{2}$.

4. For $m = 3$ and $n = 5$:

    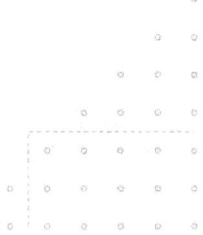

    Note that $\binom{3}{2} = 3$ is the 2nd triangular number, $\binom{5}{2} = 10$ is the 4th triangular number, and $\binom{8}{2}$ is the 7th triangular number.

1.2.4 Non-combinatorial Proofs in Combinatorics

In this section we have stressed a particular style of proof unique to combinatorics: *combinatorial proof*. As suggested a few place above, there are often other interesting proofs of combinatorial results that don't use this technique. We close this section with a few more examples of particularly elegant, non-combinatorial proofs.

We start with a well known theorem about the relationship between square and triangular numbers.

Theorem 1.2.17
$n^2 = \binom{n}{2} + \binom{n+1}{2}$.

Proof. The key insight here is that $\binom{n+1}{2} = T_n$, the nth triangular number. With this, there is an easy "picture proof." Here is the picture for $n = 4$:

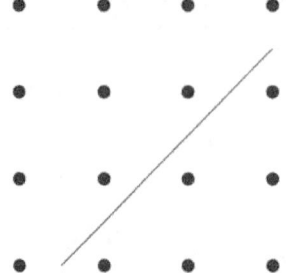

Using the factorial formula for $\binom{n}{k}$ is also straight forward, although dull. ∎

This result is surprisingly useful.

Example 1.2.18
Prove $1^2 + 2^2 + 3^2 + \cdots + n^2 = \binom{n+2}{3} + \binom{n+1}{3}$.

Solution. The sum can be recharacterized using $n^2 = \binom{n}{2} + \binom{n+1}{2}$ as follows:

$$1^2 + \left[\binom{2}{2} + \binom{3}{2}\right] + \left[\binom{3}{2} + \binom{4}{2}\right] + \cdots + \left[\binom{n}{2} + \binom{n+1}{2}\right]$$

$$= \left[\binom{2}{2} + \binom{3}{2} + \binom{4}{2} + \cdots + \binom{n+1}{2}\right] + \left[\binom{2}{2} + \binom{3}{2} + \cdots + \binom{n}{2}\right]$$

$$= \binom{n+2}{3} + \binom{n+1}{3}$$

using the Hockey Stick Theorem twice.

Exercise 40
Use Theorem 1.2.17 to prove $1^2 + 3^2 + 5^2 + \cdots + n^2 = \binom{n+2}{3}$ for odd n.

Exercise 41
The following identity has its roots in the problem of enumerating the number of squares in an n by $n + 1$ checkerboard.

$$1 \cdot n^2 + 2(n-1)^2 + 3(n-2)^2 + \cdots + n \cdot 1^2 = \binom{n+2}{4} + \binom{n+3}{4}.$$

Prove this. [Hint]

1.3 Counting with Equivalence

So far we have been able to solve our counting problems simply by use of the sum and product principles, together with the fact that the number of k-element subsets of an n-element set is $\binom{n}{k}$ (the values of which can be found in Pascal's triangle). This means that we have always *built up* the set of outcomes by combining smaller sets of outcomes combined in various ways.

It is often useful to go the other way. We could purposely over count the number of outcomes, but in a way that we can recover the correct set.

1.3.1 The Quotient Principle

In Exercise 29, you proved the algebraic formula $\binom{n}{k} = \frac{n!}{(n-k)!k!}$. However, this was done rather indirectly. First you proved the identity $P(n,k) = \binom{n}{k}k!$, then wrote $P(n,k) = \frac{n!}{(n-k)!}$, and finally solved for $\binom{n}{k}$.

This approach has the advantage of illustrating the connection between combinations and permutations, but the combinatorial argument that both sides of the identity are equal asked you to count the number of *permutations* in two ways. It is instructive to think of this equation in terms of *combinations* instead. Let's start with a concrete example.

> **Activity 42**
>
> Let's count the number of subsets of $\{a, b, c, d, e\}$ of size 3. Of course we already know the answer should be $\binom{5}{3}$, which we can see in Pascal's triangle is equal to 10. What might you try if you didn't already know this?
>
> (a) You might first guess that the answer is $5 \cdot 4 \cdot 3$ since there are 5 choices for which element you put in your subset first, then 4 choices for the next element, and 3 choices for the last element. Write down all 60 of the outcomes you get by counting the "subsets" this way.[1]
>
> (b) One of the subsets we are actually interested in is $\{a, c, d\}$. How many of the outcomes you listed above correspond to this set? How many outcomes correspond to the set $\{c, d, e\}$?
>
> (c) Explain why every subset corresponds to the same number of permutations. Then use this to count the total number of subsets correctly.

Let's look carefully at what we did above. We had a set of permutations (all 60 of the 3-permutations of the set $\{a, b, c, d, e\}$). Then we noticed that some of these permutations corresponded the the same subset. Saying that these permutations were related like this feels like defining a equivalence relation.

---

[1] This might seem like a lot of busy work, but your efforts will be rewarded.

Activity 43

Refer back to your list of 60 3-permutations of $\{a, b, c, d, e\}$.

(a) Define an **equivalence relation** on the permutations you listed so that permutations that "correspond" to the same subset are equivalent. That is, give a rule that specifies when two permutations are "equivalent". [Hint]

(b) In any set S, if you have an equivalence relation $\sim$, you can **partition** S into **equivalence classes**: sets of elements that are equivalent under $\sim$ (i.e., sets of the form $\{x \in S : x \sim a\}$ for a particular element a).

Write out the equivalence classes generated by the equivalence relation you gave above. Explain why these all have the same size. How many equivalence classes do you have (and how does this relate to the fact that they all have the same size)?

(c) Find a bijection between the set of equivalence classes and the set of subsets of $\{a, b, c, d, e\}$. Why is this important? [Hint]

This suggests a general approach to solving a counting problem. Say we want to count the number of elements in a set S. Suppose we can count the size of some set T, perhaps containing many more elements than we are actually interested in counting. Then define an equivalence relation $\sim$ on T so that there is a bijection between S and the set of equivalence classes $T/\sim$. Then if we know the size of $T/\sim$, we have the size of S.

Look at the notation $T/\sim$ that we used to denote the set of equivalence classes of T under $\sim$. This set is sometimes called the **quotient set** or **quotient space** of T by $\sim$. This is a good name, as you are dividing up a set into parts. It is also a good name because sometimes we can conclude

$$|T/\sim| = |T|/|A|$$

where A is one of the equivalence classes under $\sim$. When can you do this?

Note this is exactly what we did when we counted subsets: we divided the number of permutations by the number of permutations equivalent to each other. That worked precisely because *all the equivalence classes had the same size*!

This justifies the **quotient principle** we now state.

The Quotient Principle.

If we partition a set of size p into q blocks, each of size r, then $q = p/r$.

Like the product principle, the quotient principle is really just a statement about how division works. Here are a few examples of how you can use it in counting. In each of the following activities, make sure you can say exactly how the quotient principle could be used.

Exercise 44
In how many ways may n people sit around a round table? (Assume that when people are sitting around a round table, all that really matters is who is to each person's right. For example, if we can get one arrangement of people around the table from another by having everyone get up and move to the right one place and sit back down, we get an equivalent arrangement of people. Notice that you can get a list from a seating arrangement by marking a place at the table, and then listing the people at the table, starting at that place and moving around to the right.) There are at least two different ways of doing this problem. Try to find them both, especially the one that uses the quotient principle. [Hint]

Exercise 45
In how many ways may we string n distinct beads on a necklace without a clasp? (Perhaps we make the necklace by stringing the beads on a string, and then carefully gluing the two ends of the string together so that the joint can't be seen. Assume someone can pick up the necklace, move it around in space and put it back down, giving an apparently different way of stringing the beads that is equivalent to the first.) [Hint]

Exercise 46
We first gave this problem as Exercise 25. Now we have several ways to approach the problem. A tennis club has $2n$ members. We want to pair up the members by twos for singles matches.

(a) In how many ways may we pair up all the members of the club? Give at least two solutions different from the one you gave in Exercise 25. (You may not have done Exercise 25. In that case, see if you can find three solutions.) [Hint]

(b) Suppose that in addition to specifying who plays whom, for each pairing we say who serves first. Now in how many ways may we specify our pairs? Try to find as many solutions as you can. [Hint]

Exercise 47
In how many ways may we attach two identical red beads and two identical blue beads to the corners of a square (with one bead per corner) free to move around in (three-dimensional) space? [Hint]

Example 1.3.1
We have used the quotient principle to explain the formula $\binom{n}{k} = \frac{n!}{(n-k)!k!}$ by thinking of this as $\binom{n}{k} = \frac{P(n,k)}{k!}$. What if we don't involve $P(n,k)$ at all?

Describe a set of outcomes that has size $n!$ that can be partitioned into blocks of size $(n-k)!k!$ so that each block corresponds to something $\binom{n}{k}$ counts.

Solution. Suppose you have k red balls and $n-k$ blue balls. Each ball has a different number printed on it. Thus there are $n!$ different ways to arrange the n balls in a line. But what if we only care about the color pattern the balls make (perhaps the numbers have been printed in invisible ink and have now vanished). We can arrange these permutations of all n balls into blocks, where two permutations are in the same block if and only if they have the same sequence of colors (RRBRB...). There are $(n-k)!k!$ permutations in each block, since the red balls can be arranged in $k!$ ways and the blue balls can be arranged in $(n-k)!$ ways.

All this says that the number of two-color patterns of length n including k red balls is $\frac{n!}{(n-k)!k!}$. But of course the answer is also $\binom{n}{k}$ because of the n positions, we must choose k positions to put the identical red balls.

Exercise 48

How many anagrams of the word "anagram" are there? (An anagram is a rearrangement of *all* of the letters of a word.) [Hint]

1.3.2 When does Order Matter?

One way to distinguish combinations from permutations is by asking whether "order matters". Both $\binom{n}{k}$ and $P(n,k)$ count the number of ways to select k objects from n objects without repeats, but you use the combination $\binom{n}{k}$ when "order doesn't matter" and the permutation $P(n,k)$ when "order matters". Despite the presence of the scare quotes, this is not a false statement, but we must understand what we mean when we say "order matters".

Activity 49

Each counting question below asks for two answers. Decide which answer is a combination and which is a permutation, and why that makes sense.

(a) An ice-cream shop offers 31 flavors. How many 3-scoop ice-cream cones are possible, assuming each scoop must be a different flavor? How many 3-scoop milkshakes are possible, assuming each scoop must be a different flavor? [Hint]

(b) How many 5-digit numbers are there with distinct, non-zero digits for which the digits must be increasing? How many are there for which the digits can come in any order? [Hint]

(c) How many injective functions $f : [k] \to [n]$ are there all together? How many injective functions $f : [k] \to [n]$ are there that are (strictly) increasing? [Hint]

Exercise 50
The number of n-bit strings of weight k is $\binom{n}{k}$, so a combination. But in determining one bit string from another, all that matters is the order in which the k 1's and $n-k$ 0's appear. So does order matter? In what sense does it not? [Hint]

Exercise 51
Write a clear sentence or two saying specifically what we mean when we say "order matters" to distinguish between combinations and permutations. [Hint]

Understanding the role of order in distinguishing outcomes often suggests the use of the quotient principle. You might say we arrive at combinations by counting permutations and "modding" out by the order. Each block that corresponds to a single combination is a group of permutations that are only different because of their order. Let's see if we can apply this to other counting questions.

Activity 52
The ice cream shop is down to only 3 flavors. If you wanted a 3-scoop cone or a 3-scoop shake, made without repeated flavors, there would only be 6 cones possible and only 1 shake. But what if you allowed repeated flavors?

(a) How many 3-scoop cones are possible? [Hint]

(b) How many 3-scoop shakes are there? Write all of them down. [Hint]

(c) Why doesn't the quotient principle apply here? What goes wrong?

List out all 3-scoop cones and form the equivalence classes of shakes to see the issue.

The previous activity illustrates that you cannot always simply apply the quotient principle to eliminate order mattering. What we are after in the 3-scoop shakes with possibly repeated flavors is a **multiset**, which is just like a set only we allow for elements to be in the set multiple times, but we still do not care in what order the elements are listed. So an example of an 4 element multiset of $[7]$ is $\{1, 3, 3, 6\}$, which is the same multiset as $\{1, 6, 3, 3\}$ but different from the multiset $\{1, 3, 6\}$ (in particular, this multiset only has three elements).

What we tried to do above was to count multisets by first counting sequences with possibly repeated elements, and then divide out by the number of sequences that just differ by the arrangement of a particular multiset. But this didn't work, because the size of each block was not the same.

However, we would still like to be able to count the number of multisets. How can we do this? Is it an application of the quotient principle after all? To start, let's establish some notation. The number of k-element subsets of $[n]$ was given by n choose k (written $\binom{n}{k}$), so we will say that the number of k-element *multisets* of $[n]$ will be n **multichoose** k, and write $\left(\!\binom{n}{k}\!\right)$. From what we have seen above, we can say,

$$\left(\binom{n}{k}\right) \neq \frac{n^k}{k!}$$

despite the obvious allure of this possibility.

Our question now is, what *should* be in the numerator if not n^k? First let's get more comfortable with thinking of multisets as a model for counting questions.

Activity 53

Explain why each of the following counting problems have an answer in the form $\left(\binom{n}{k}\right)$ (say what n and k should be). That is, show how each of them is counting the number of k-element multisets of $[n]$ (or a different set of size n).

(a) If you have an unlimited supply of pennies, nickels, dimes and quarters, how many handfuls of 7 coins can you grab?

(b) You roll 10 regular 6-sided dice (because you enjoy playing two games of Yahtzee at once). Assuming the dice are indistinguishable, how many different outcomes are possible?

(c) How many functions $f : [5] \to [7]$ are non-decreasing? For example, one such function is $f = \begin{pmatrix} 1 & 2 & 3 & 4 & 5 \\ 2 & 5 & 5 & 6 & 7 \end{pmatrix}$.

Activity 54

Explain why each of the following counting problems are also answered in the form $\left(\binom{n}{k}\right)$ (and say what n and k are). You should be able to explain why all of these are equivalent to each other, and then find a bijection from any of them to the set of k-element multisets of $[n]$.

(a) How many non-negative integer solutions are there to the equation $x_1 + x_2 + x_3 + x_4 = 10$?

(b) How many ways are there to distribute 7 identical cookies to 5 kids? Some kids might not get any cookies, but you should distribute all 7 cookies.

(c) How many ways can you put 10 identical books on the 4 shelves of a bookcase? Assume each shelf could hold up to 10 books, and that all the books are always shoved all the way to the left (so we are not worried about where the books are on an individual shelf, just which shelf they go on).

The problems in Activity 54 are sometimes referred to as "distribution problems", since we are asking for the number of ways to distribute some objects to some recipients. This is a useful model for the other counting techniques we have seen as well. For example, a permutation $P(n,k)$ gives the number of ways to distribute n distinct objects to k distinct recipients, so that each recipient gets exactly one object.

The claim implicit in Activity 54 is that the number of ways to distribute n identical objects to k distinct recipients (now allowing each recipient to receive any number of objects, including zero) is $\left(\!\binom{n}{k}\!\right)$. If this is the case, then we should be able to think of the set of k-element multisets of $[n]$ as a distribution.

Activity 55
Consider 3-element multisets of $[5]$. One of these is $\{1, 1, 1\}$, and another is $\{2, 4, 5\}$. How are these two multisets like distributing three objects among 5 recipients? What are the objects that you are distributing?

Now let's consider a related distribution problem, where in some way, order now matters. In fact, how is it that saying the distribution problems as in Activity 54 are ones for which order doesn't matter? Look back at our standard example of order mattering or not:

Activity 56

(a) If $P(n, k)$ counts the number of ways to distribute n distinct objects to k distinct recipients in which each recipient receives exactly one object (do you see why this is?), then what distribution does $\binom{n}{k}$ count? Explain what is playing the role of "order mattering" in a distribution problem.

(b) We claim that counting the number of ways to distribute n identical books to k distinct shelves is counted by $\left(\!\binom{n}{k}\!\right)$, and this is a situation in which order does not matter (since we are counting multisets). What would the distribution question be if order did matter? Explain why the answer to that question is NOT n^k. (What distribution does *that* expression count?)

Consider the case where the books are distinct instead of identical. Note also that in the next few activities, the usage of the variables k and n are reversed from above. As you work through these next activities, think about why that choice was made.

Activity 57
Suppose we wish to place k distinct books onto the shelves of a bookcase with n shelves. For simplicity, assume for now that all of the books would fit on any of the shelves. Also, let's imagine pushing the books on a shelf as far to the left as we can, so that we are only thinking about how the books sit relative to each other, not about the exact places where we put the books. Since the books are distinct, we can think of the first book, the second book and so on.

(a) How many places are there where we can place the first book?

(b) When we place the second book, if we decide to place it on the shelf that already has a book, does it matter if we place it to the left or right of the book that is already there?

   (c) How many places are there where we can place the second book?
   [Hint]

   (d) Once we have $i - 1$ books placed, if we want to place book i on a shelf that already has some books, is sliding it in to the left of all the books already there different from placing it to the right of all the books already or between two books already there?

   (e) In how many ways may we place the ith book into the bookcase?
   [Hint]

   (f) In how many ways may we place all the books?

The assignment of which books go to which shelves of a bookcase is simply a function from the books to the shelves. But a function does not determine which book sits to the left of which others on the shelf, and this information is part of how the books are arranged on the shelves. In other words, the order in which the shelves receive their books matters. Our function must thus assign an ordered list of books to each shelf. We will call such a function an ordered function. More precisely, an **ordered function** from a set S to a set T is a function that assigns an (ordered) list of elements of S to some, but not necessarily all, elements of T in such a way that each element of S appears on one and only one of the lists.[2]

We often think of functions as a rule that assigns an output to each input. However, we could equally well specify a function by giving the complete inverse image of each element of the codomain. That is, to define a (non-ordered) function $f : S \to T$, we could give the set $f^{-1}(t) = \{s \in S : f(s) = t\}$ for each $t \in T$. The set of elements sent to t is a *set* for a function, but for an ordered function, it is a *sequence*.

Theorem 1.3.2
The number of ordered functions from a k-element set to an n-element set is

$$\prod_{i=1}^{k}(n - 1 + i) = \frac{(n - 1 + k)!}{(n - 1)!} = P(n - 1 + k, k).$$

Activity 58

   (a) In some sort of manufacturing mishap, your set of magnetic letters contains only the first 7 letters of the alphabet, and for some reason

---

[2] The phrase ordered function is not a standard one, because there is as yet no standard name for the result of an ordered distribution problem.

5 identical exclamation marks. How many ways can you arrange all 12 magnets in a single line? (one such line is BG!AD!!!FEC!).

(b) What does this question have to do with placing 7 distinct books on 6 shelves? [Hint]

(c) Oh no! Your 5 year old left the 7 magnetic letters in the oven too long and now they are all identical blobs of plastic. How many strings of the 12 magnets can you make now? [Hint]

(d) What sort of distribution problem does the previous task correspond to? Write a question about books and shelves that has the same answer.

AHA! The previous activity suggests that to count multisets, which is the same as counting distributions of identical objects to distinct recipients, we should apply the quotient principle to ordered functions.

What should the equivalence relation be? Think about bookshelves. An ordered function is a distribution of distinct books to distinct shelves. A multiset is a distribution of identical books to distinct shelves. So we want two ordered functions to be equivalent if their corresponding shelves contain the same number of books (we no longer care what those books are, just how many there are).

Now, given any ordered function, how many ordered functions are in its equivalence class? We can get any other element in the class by permuting the books (but keeping the numbers in each shelf constant). There are k books, so there are $k!$ ways to permute them. This is the same for *every* ordered function, so the equivalence classes all have the same size.

Applying the quotient principle, we arrive at the following.

Theorem 1.3.3
The number of k-element multisets of $[n]$ is
$$\left(\!\!\binom{n}{k}\!\!\right) = \frac{P(n-1+k, k)}{k!} = \binom{n-1+k}{k}$$

It is interesting that the number of k-element multisets of $[n]$ can be expressed as a binomial coefficient. This suggests there should be a way to describe a multiset as choosing k things from a collection of $n - 1 + k$ things. Or perhaps thinking about each multiset as a $(n - 1 + k)$-bit string of weight k.

There is a standard interpretation for this. We call this approach **stars and bars** (others use *balls and wall* or *dashes and slashes*; the important thing is that it rhymes). The idea is that we represent each multiset as a string of k "stars" and $n - 1$ "bars". These are really just bit strings in disguise, so the binomial coefficient becomes clear. The only question is how do multisets correspond to the strings?

Activity 59
Consider 4-element multisets of [3]. Using the formula above, there are $\binom{6}{4} = 15$ of these. There are also $\binom{6}{4}$ stars and bars strings using 4 stars and 2 bars.

(a) Write out all 15 multisets and all 15 stars and bars strings. Find a bijection between these that makes sense. [Hint]

(b) Why does it make sense that there are only 2 bars in these diagrams, but 3 elements that can go into the multiset?

(c) What would go wrong if we used 3 bars and 3 stars (that is, confused n and k)? Why does $*||**|$ not make sense as a multiset here?

Thinking of these distribution problems as "stars and bars problems" can make solving them very quick. The key step is deciding what part of the problem the stars correspond to, and what the bars correspond do. The remaining exercises here should give you some practice with this.

Exercise 60
Each of the counting problems below can be solved with stars and bars. For each, say what outcome the string

$$***|*||**|$$

represents, if there are the correct number of stars and bars for the problem. Otherwise, say why the diagram does not represent any outcome, and what a correct diagram would look like.

1. How many ways are there to select a handful of 6 jellybeans from a jar that contains 5 different flavors?

2. How many ways can you distribute 5 identical lollipops to 6 kids?

3. How many 6-letter words can you make using the 5 vowels in increasing order?

4. How many solutions are there to the equation $x_1 + x_2 + x_3 + x_4 = 6$.

Exercise 61
In Activity 53 and Activity 54 you considered some counting questions we claimed could be expressed as multisets. Now solve each of these using stars and bars and the formula in Theorem 1.3.3. Be sure to clearly state which part of the problem corresponds to the stars and which to the bars, and why this makes sense.

(a) If you have an unlimited supply of pennies, nickels, dimes and quarters, how many handfuls of 7 coins can you grab?

(b) How many non-negative integer solutions are there to the equation $x_1 + x_2 + x_3 + x_4 = 10$?

(c) You roll 10 regular 6-sided dice (because you enjoy playing two games of Yahtzee at once). Assuming the dice are indistinguishable, how many different outcomes are possible?

(d) How many ways are there to distribute 7 identical cookies to 5 kids? Some kids might not get any cookies, but you should distribute all 7 cookies.

(e) How many functions $f : [5] \to [7]$ are non-decreasing? For example, one such function is $f = \begin{pmatrix} 1 & 2 & 3 & 4 & 5 \\ 2 & 5 & 5 & 6 & 7 \end{pmatrix}$.

(f) How many ways can you put 10 identical books on the 4 shelves of a bookcase? Assume each shelf could hold up to 10 books, and that all the books are always shoved all the way to the left (so we are not worried about where the books are on an individual shelf, just which shelf they go on).

1.3.3 More Distribution Problems

Both multisets and ordered functions allowed some of the recipients to receive nothing (or for some elements of $[n]$ to not be included in the multiset). What if we didn't want to allow this?

Activity 62
Suppose we wish to place the books in Activity 57 (satisfying the assumptions we made there) so that each shelf gets at least one book. Now in how many ways may we place the books? (Hint: how can you make sure that each shelf gets at least one book before you start the process described in Activity 57?) [Hint]

The previous activity is an example of what we might call an **ordered surjection**. Just like we did with ordered functions, we can apply the quotient principle to count distributions where the objects being distributed are no longer distinct.

Exercise 63
In how many ways may we put k identical books onto n shelves if each shelf must get at least one book? [Hint]

Example 1.3.4
A **composition** of the integer k into n parts is a list of n positive integers that add to k. How many compositions are there of an integer k into n parts?

Solution. There is a bijection between compositions of k into n parts and arrangements of k identical books on n shelves so that each shelf gets a book. Namely, the number of books on shelf i is the ith element of the list. Thus the number of compositions of k into n parts is $\binom{k-1}{n-1}$.

Example 1.3.5
The answer in Example 1.3.4 can be expressed as a binomial coefficient. This means it should be possible to interpret a composition as a subset of some set. Find a bijection between compositions of k into n parts and certain subsets of some set. Explain explicitly how to get the composition from the subset and the subset from the composition.

Solution. If we line up k identical books, there are $k-1$ places in between two books. If we choose $n-1$ of these places and slip dividers into those places, then we have a first clump of books, a second clump of books, and so on. The ith element of our list is the number of books in the ith clump. Clearly using books is irrelevant; we could line up any k identical objects and make the same argument. Our bijection is between compositions and $(n-1)$-element subsets of the set of $k-1$ spaces between our objects.

Exercise 64
Explain the connection between compositions of k into n parts and the problem of distributing k identical objects to n recipients so that each recipient gets at least one.

Exercise 65
The previous exercise suggests that you can represent compositions as stars and bars strings as well.

(a) Consider the composition of 5 into 3 parts: $1 + 3 + 1$. This is like giving 1 apple to the first kid, 3 to the second, and 1 to the last. What stars and bars string would you use for this?

(b) Explain why $*||****$ does not represent any composition. What must be true of a stars and bars string for it to correspond to a composition?

(c) Use your classification of valid stars and bars strings to justify that there are $\binom{k-1}{n-1}$ compositions of k into n parts.

So far, we have only considered distribution problems in which the recipients are distinct. There are plenty of situations where we would want to consider recipients identical. For example, a problem we will take up in Chapter 2 is how many ways we can partition an integer. For example, we could partition 5 as $4 + 1$ or $3 + 1 + 1$, but we would not want to also count $1 + 4$ or $1 + 3 + 1$ (doing so would be creating a *composition*, which we have already considered). We can think of the integer partition problem as distributing the 5 units that make up 5 into some number of identical bins.

Here is another example, which we already have the tools to address.

> **Exercise 66**
> In how many ways may we stack k distinct books into n identical boxes so that there is a stack in every box? There are two distinct ways to answer this question. Find them both. [Hint]

We can think of stacking books into identical boxes as partitioning the books and then ordering the blocks of the partition. This turns out not to be a useful computational way of visualizing the problem because the number of ways to order the books in the various stacks depends on the sizes of the stacks and not just the number of stacks. However this way of thinking actually led to the first hint in Exercise 66. Instead of dividing a set up into non-overlapping parts, we may think of dividing a *permutation* (thought of as a list) of our k objects up into n ordered blocks. We will say that a set of ordered lists of elements of a set S is a **broken permutation** of S if each element of S is in one and only one of these lists.[3] The number of broken permutations of a k-element set with n blocks is denoted by $L(k,n)$. The number $L(k,n)$ is called a **Lah Number** and, from our solution to Exercise 66, is equal to $k!\binom{k-1}{n-1}/n!$.

The Lah numbers are the solution to the question "In how many ways may we distribute k distinct objects to n identical recipients if order matters and each recipient must get at least one?"

---

[3] The phrase broken permutation is not standard, because there is no standard name for the solution to this kind of distribution problem.

1.4 Counting with Recursion

What happens when you get stuck on a counting problem? It is quite easy to miss the clever way of thinking about breaking down a task that leads to a solution. Here is an example.

Your favorite milk-shake diner has 11 stools along the bar. Patrons of the diner have a rule that no two adjacent stools can be simultaneously occupied. In how many ways can the stools be occupied by any number of patrons (including no patrons at all, leaving all stools unoccupied)?

A first attempt might be to start with stool 1 and realize that it can be sat on or not. So there are two choices for stool 1. But now you might have two choices for stool 2 (if stool 1 was not sat on) or only once choice (to keep stool 2 empty, if stool 1 was in use). Okay, so the multiplicative principle doesn't help.

Say you can't think of anything else to try. The problem seems just too hard. So make it easier: 11 is way too big of a number; 1 would be much better. How many ways could a row of 1 stool be occupied? Let's call this number S_1. Maybe then we also try the version with 2 and 3 stools.

> **Activity 67**
> List all stool seating charts with 1, 2, and 3 stools. Let the number of seating charts be called S_1, S_2, and S_3 respectively. Do you notice a pattern among these numbers? [Hint]

We get a sequence $(S_n)_{n\geq 1}$ which gives the number of seating charts.

The sequence we get is $2, 3, 5, 8, 13, \ldots$. It *appears* that this sequence satisfies a familiar recurrence relation of the **Fibonacci sequence**: $S_n = S_{n-1} + S_{n-2}$. But unlike the Fibonacci sequence, here the initial terms are $S_1 = 2$ and $S_2 = 3$. Is this appearance deceiving? Can we explain why this sequence should satisfy this recurrence?

> **Activity 68**
> Explain why S_n satisfies the recurrence $S_n = S_{n-1} + S_{n-2}$. You will need to consider seating charts with n, $n-1$ and $n-2$ stools. Show how you can "find" the seating charts for $n-1$ and $n-2$ stools among those for n stools (perhaps using some sort of bijection). [Hint]

Once we have a proof that our sequence really does give the answers to all the versions of the counting question we are interested in, then we can start investigating properties of this sequence. Sometimes, we might even be able to find a closed formula for the nth term of the sequence, so we can compute values without relying on the recurrence.

1.4.1 Finding Recurrence Relations

Before investigating properties of sequences given by recurrence relations, we must be able to find those sequences and the recurrence relations. The following activities give some practice with moving from a counting question to a recurrence relation for the sequence of answers. You might be able to answer the counting questions with what we have already discovered, but do not do this. The point of these is to start thinking about these counting problems recursively.

First, here is an example of what we are looking for.

> **Example 1.4.1**
>
> Let a_n be the number of subsets of $[n]$. Find a recurrence relation for the sequence $(a_n)_{n\geq 0}$. Prove the recurrence relation is correct.
>
> **Solution.** Start by enumerating the sequence $(a_n)_{n\geq 0}$. There is only 1 subset of the empty set (so $a_0 = 1$). There are two subsets of $[1]$, and four subsets of $[2]$. So the sequence looks like $1, 2, 4, 8, \ldots$. It *appears* that the recurrence relation should be $a_n = 2a_{n-1}$ with initial condition $a_0 = 1$.
>
> To prove this is correct, we *must* reference the things we are counting (saying that the recurrence agrees with the finitely many terms we have listed is NOT enough). Consider the a_{n-1} different subsets of $[n-1]$. Each of these is a subset of $[n]$, in particular, a subset of $[n]$ that does not include the number n as an element. Further, to each of these sets, we can add the element n to get another distinct subset of $[n]$. So far, we have found $2a_{n-1}$ subsets of $[n]$. Actually, we have found all of them: every subset of $[n]$ either includes n or it does not.

Now here are some for you to try.

> **Exercise 69**
>
> Let k_n be the number of handshakes that take place if all n people in a room shake hands with everyone there exactly once. Find a recurrence relation for the sequence $(k_n)_{n\geq 1}$ and prove you are correct. [Hint]

> **Exercise 70**
>
> Let b_n be the number of bijections from an n-element set to an n-element set. Find a recurrence relation for the sequence $(b_n)_{n\geq 1}$ and prove you are correct. [Hint]

> **Exercise 71**
>
> The "Towers of Hanoi" puzzle has three rods rising from a rectangular base with n rings of different sizes stacked in decreasing order of size on one rod. A legal move consists of moving a ring from one rod to another so that it does not land on top of a smaller ring. If m_n is the number of moves required to move all the rings from the initial rod to another rod that you choose, give a recurrence for m_n.
>
>
>
> [Hint]

Exercise 72

We draw n mutually intersecting circles in the plane so that each one crosses each other one exactly twice and no three intersect in the same point. (As examples, think of Venn diagrams with two or three mutually intersecting sets.) Find a recurrence for the number r_n of regions into which the plane is divided by n circles. (One circle divides the plane into two regions, the inside and the outside.) Find the number of regions with n circles. For what values of n can you draw a Venn diagram showing all the possible intersections of n sets using circles to represent each of the sets? [Hint]

The sequences above all have relatively simple recurrence relations. In particular, the recursions all make reference only to one previous term. This need not be the case, as we saw when counting bar stool seating charts and as the next few problems demonstrate.

Exercise 73

Suppose you have a large number of 1×1 squares and 1×2 dominoes. You wish to use these to tile a $1 \times n$ path. For $n = 4$, some examples of different paths are $ssss$, ssd, sds, dss, dd, using s for square and d for domino (in fact, these are all 5 of the possible paths).

(a) Let a_n be the number of ways to tile the $1 \times n$ path. Find a recurrence relation for the sequence $(a_n)_{n \geq 1}$, and prove you are correct. [Hint]

(b) Now suppose your squares come in 3 colors and the dominoes come in 4 colors. Let b_n count the number of paths you can tile. Find and justify a recurrence relation for $(b_n)_{n \geq 1}$.

Exercise 74

Consider a convex n-gon with labeled vertices. In how many ways can diagonals be inserted so as to decompose the n-gon into triangles? For $n = 5$ the five figures are shown in Figure 1.4.2.

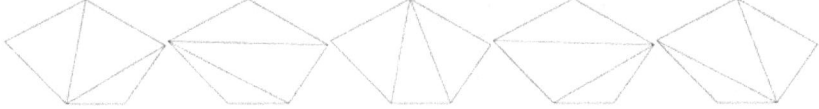

Figure 1.4.2 The five triangulations of a convex pentagon.

Let t_n be the number of triangulations. Find a recurrence relation for $(t_n)_{n \geq 3}$. Justify your answer. [Hint]

We will study the recurrence relation in Exercise 74 again in Section 1.5. We mention it here because it is a nice example of how sometimes recurrence relations use all the previous terms of the sequence. Sometimes we can find a simpler recurrence relation that captures the same sequence.

Exercise 75

Consider the sequence $(a_n)_{n\geq 1}$ which satisfies the recurrence relation $a_n = \sum_{i=1}^{n-1} a_i$. That is, each term of the sequence is the sum of *all* previous terms in the sequence.

Find a recurrence relation in terms of only the previous term for this sequence. Prove that you are correct. [Hint]

1.4.2 Using Recurrence Relations

While it is often easier to find a recurrence relation governing a sequence, if we wish to use sequences to solve counting problems, we would really like to be able to find a closed formula for the terms in a sequence. If this is not possible, we can still use the recurrence relation to learn something about how the sequence behaves.

We will start by consider some simple classes of recurrence relations for which we can find closed formulas.

Activity 76

Let $(a_n)_{n\leq 0}$ satisfy $a_n = a_{n-1} + 3$ with $a_0 = 2$. For example, a_n might give the number of push-ups you can do n days into your training, assuming you can do 3 more push-ups each day, if you could do 2 push-ups before you started training. Find a closed formula for a_n and justify your answer. Show how the recurrence relation is used. [Hint]

Activity 77

Let $(a_n)_{n\leq 0}$ satisfy $a_n = 3a_{n-1}$ with $a_0 = 2$. For example, a_n might give the number of n-scoop ice-cream cones you can make from 3 available flavors in one of two kinds of cones. Find a closed formula for a_n and justify your answer. Show how the recurrence relation is used. [Hint]

The sequence in Activity 76 is an example of an **arithmetic sequence** (or arithmetic progression) in that the difference between terms is constant. The sequence in Activity 77 is an example of a **geometric sequence** (or geometric progression) because the *ratio* between terms is constant. Of course there is nothing special about 2 and 3 in the activities, so you should now see how to find the closed formula for any arithmetic or geometric sequence.

The techniques used to get those closed formulas are also worth pointing out. The method suggested in the hint to Activity 76 is called **telescoping** and the method suggested for Activity 77 is called **iteration**.

Most sequences are not arithmetic, but it is still helpful to look at the difference between terms. Often if we know something about the sequence of differences, we can deduce something about the original sequence. We can also think about this as starting with the sequence of differences.

Given any sequence $(a_n)_{n\geq 1}$, we can form the **sequence of partial sums** $(b_n)_{n\geq 1}$ given by $b_n = \sum_{i=1}^{n} a_n$. That is, b_n tells you what you get if you add up the first n terms of (a_n).

Note that (b_n) satisfies the recurrence relation $b_n = b_{n-1} + a_n$, or equivalently $b_n - b_{n-1} = a_n$. So (a_n) is the sequence of differences of its sequence of partial sums!

Example 1.4.3

If $(a_n)_{n\geq 1}$ is the familiar sequence $1, 2, 3, 4, \ldots,$, what is the sequence of partial sums?

Solution. The sequence of partial sums is $1, 3, 6, 10, 15, \ldots$; the **triangular numbers** (so called because you can arrange those numbers of dots into equilateral triangles). A closed formula for these is $T_n = \frac{n(n+1)}{2}$, which can be found as follows:

$$
\begin{array}{rccccccccc}
T_n = & 1 & + & 2 & + \cdots + & (n-1) & + & n \\
+\ T_n = & n & + & (n-1) & + \cdots + & 2 & + & 1 \\
\hline
2T_n = & (n+1) & + & (n+1) & + \cdots + & (n+1) & + & (n+1)
\end{array}
$$

The right hand side contains n summands, all identical, so $2T_n = n(n+1)$. Divide by 2!

Notice also that these triangular numbers appear as the third diagonal of Pascal's triangle. That is, $T_n = \binom{n+1}{2}$. Using the factorial formula for $\binom{n+1}{2}$ confirms this.

Exercise 78

(a) If $(a_n)_{n\geq 1}$ is the sequence $1, 3, 5, 7, 9, \ldots$, find the sequence of partial sums (b_n).

(b) If $a_n = \binom{n}{2}$, find the sequence of partial sums (b_n).

(c) Consider the Fibonacci sequence $(F_n)_{n\geq 1}$, starting $1, 1, 2, 3, 5, \ldots$, and satisfying the recurrence relation $F_n = F_{n-1} + F_{n-2}$. Find the sequence of partial sums. How do the terms relate to the original sequence?

The sequences in the previous problem worked out nicely, but what can we say in general? Here are a few things.

Activity 79

Let $(a_n)_{n\geq 1}$ be any arithmetic sequence. Say $a_n = dn + c$ (so d is the common difference and $c = a_1 - d = a_0$, if a_0 were part of the sequence). What can we say about the closed formula for $b_n = \sum_{i=1}^{n} a_i$?

(a) As an example, what is $2 + 5 + 8 + 11 + \cdots + 470$? You might call this sum S and calculate:

$$
\begin{array}{rccccccccc}
S = & 2 & + & 5 & + & 8 & + \cdots + & 467 & + & 470 \\
+\ S = & 470 & + & 467 & + & 464 & + \cdots + & 5 & + & 2 \\
\hline
2S = & 472 & + & 472 & + & 472 & + \cdots + & 472 & + & 472
\end{array}
$$

Why is that helpful? What is the sum? [Hint]

(b) Generalize the previous computation to find the closed formula for $b_n = \sum_{i=1}^{n} a_i$.

(c) What is special about sums of arithmetic sequences that allows you to do this computation? Explain.

The previous activity shows that the sequence of partial sums of an arithmetic sequence is always a quadratic sequence. Conversely, any quadratic sequence should arise in this way.

Activity 80
Given a sequence $(b_n)_{n \geq 1}$ with a quadratic closed formula, say $b_n = an^2 + bn + c$, what will the closed formula for the sequence of differences be? That is, what is the closed formula for $a_n = b_n - b_{n-1}$? Why does this show that every quadratic sequence is the sequence of partial sums for some arithmetic sequence?

The above observations can be generalized: a cubic sequence will have quadratic difference, a quartic sequence will have cubic differences, and so on. This allows us to guess at the *form* of a closed formula for a polynomial sequence by taking successive differences. If the "third differences" of a sequence are constant, then we would guess we could find a cubic polynomial. Given the first 4 terms, we can fit the cubic to them by setting up a system of 4 equations and 4 unknowns.

Sequences of partial sums are also useful for geometric sequences. The following technique will likely be familiar from calculus when you worked with geometric series.

Activity 81
What is $3 + 6 + 12 + 24 + \cdots + 12288$? Call the sum S and compute $S - 2S$.

To better see what happened in the above example, try writing it this way:

$$\begin{array}{rlr} S = & 3 + \quad 6 + 12 + 24 + \cdots + 12288 & \\ -\quad 2S = & \quad\quad 6 + 12 + 24 + \cdots + 12288 & +24576 \\ \hline -S = & 3 + \quad 0 + 0 + 0 + \cdots + 0 & -24576 \end{array}$$

Then divide both sides by -1 and we have the same result for S. The idea is, by multiplying the sum by the common ratio, each term becomes the next term. We shift over the sum to get the subtraction to mostly cancel out, leaving just the first term and new last term.

Exercise 82
Find a closed formula for $S_n = 2 + 10 + 50 + \cdots + 2 \cdot 5^n$. [Hint]

Notice that this is precisely the standard method used for converting repeating decimals to fractions.

> **Exercise 83**
> Express $N = 0.464646\ldots$ as a fraction by computing $N - 0.01N$. Why is this the same as computing an infinite geometric sum?

Here is a surprising use of these geometric sums to answer a counting question:

> **Exercise 84**
> How many license plates consist of 6 symbols, using only the three numerals 1, 2, and 3 and the four letters a, b, c, and d, so that no numeral appears after any letter? For example, "31ddac" and "12321" are acceptable license plates, but "13ba2c" is not.
>
> (a) First answer this question by considering different cases: how many of the license plates contain no numerals? How many contain one numeral, etc. Find the sum. [Hint]
>
> (b) Recognize that that the total S when written this way is the sum of a (finite) geometric sequence. Compute $S - \frac{3}{4}S$ to conclude that the answer can be expressed as $4^7 - 3^7$.
>
> (c) Generalize! What if a license plate has n symbols?

Guessing the form of a closed formula is also reasonable when we believe a sequence to be exponential.

> **Activity 85**
> Consider the recurrence relation $a_n = 5a_{n-1} - 6a_{n-2}$. Listing terms (with some reasonable initial conditions) suggests that this sequence grows exponentially.
>
> (a) Show that $a_n = 2^n$ and $a_n = 3^n$ are both solutions to the recurrence relation. That is, they are sequences for which the recurrence relation holds. [Hint]
>
> (b) Show that if α and β are any constants, $a_n = \alpha 2^n + \beta 3^n$ is also a solution to the recurrence relation.
>
> (c) Might there be another solution to the recurrence relation? Suppose $a_n = r^n$ for some non-zero constant r. Show that $r = 2$ or $r = 3$. [Hint]

This is a partial justification of the **characteristic root technique** (the polynomial whose roots are the base of the exponentials is called the **characteristic polynomial**). Making the reasonable guess that recurrence relations which give a_n as a linear combination of a_{n-1} and a_{n-2} should have exponential closed formulas allows you to find the bases of those exponentials. You can then use the initial conditions to find the constants α and β. Let's illustrate this process with an example.

Example 1.4.4
Find a closed formula for the sequence $(a_n)_{n \geq 1}$ given recursively by $a_n = 3a_{n-1} + 4a_{n-2}$ with initial conditions $a_1 = 3$ and $a_2 = 13$. (This is the sequence of colorful square/domino paths from Exercise 73.)

Solution. From the recurrence relation we have $a_n - 3a_{n-1} - 4a_{n-2} = 0$. We transform this into the characteristic equation $x^2 - 3x - 4 = 0$. This has roots $x = 4$ and $x = -1$. So we guess that the closed formula for a_n will be $a_n = \alpha 4^n + \beta(-1)^n$.

To find α and β, we use the initial conditions. When $n = 1$ we have $3 = 4\alpha - \beta$. When $n = 2$ we have $13 = 16\alpha + \beta$. It is easy to solve this system of two equations and two unknowns (simply adding them together eliminates the β). We find $\alpha = \frac{4}{5}$ and $\beta = \frac{1}{5}$.

Thus the closed formula is
$$a_n = \frac{4}{5} 4^n + \frac{1}{5}(-1)^n.$$

Exercise 86
How many $1 \times n$ paths can you make from 1×1 squares that come in 2 colors and 1×2 dominoes that come in 3 colors?

(a) First write down a recurrence relation and initial conditions for this problem.

(b) Solve the recurrence relation using the characteristic root technique.

Exercise 87
Find a closed formula for the nth Fibonacci number F_n, where $F_1 = 1 = F_2$ and $F_n = F_{n-1} + F_{n-2}$. From the form of this recurrence, the characteristic root technique should work here. (The formula you will find is called Binet's Formula.) [Hint]

There are many subtle and interesting variations on solving recurrence relations, and we will not explore them all here. To conclude though, we consider one interesting case that will show up later as well.

Activity 88
Consider the recurrence relation $a_n = 2a_{n-1} - a_{n-2}$ with initial conditions $a_0 = 1$ and $a_1 = 2$.

(a) What goes wrong when you (blindly) apply the characteristic root technique?

(b) Write out the first few terms of the sequence to guess at the closed formula. Verify it is correct.

As you saw above, something goes wrong with the characteristic root technique when the roots of the characteristic equation are repeated. One way around this is to guess that the closed formula should be

$$a_n = \alpha r^n + \beta n r^n,$$

where r is the repeated root and α and β are constants determined by the initial conditions as normal. The key here is to add an extra multiple of n to the second term.

> **Exercise 89**
> Consider the recurrence relation $a_n = 4a_{n-1} - 4a_{n-2}$.
>
> (a) Find the general solution (leaving α and β as parameters).
>
> (b) Find the solution with initial conditions $a_0 = 1$ and $a_1 = 2$.
>
> (c) Find the solution with initial conditions $a_0 = 1$ and $a_1 = 8$.

1.4.3 Exploring the Fibonacci Sequence

Even though we are thinking of sequences as a tool for solving counting questions, it is often very rewarding to step back and observe the beauty and elegance of a sequence in its own right. So to conclude this section, we will consider a few delightful observations about the most famous recursively defined sequence: the Fibonacci numbers.

For consistency, let's agree that F_n is the nth Fibonacci number, where $F_0 = 0$ and $F_1 = 1$, and for all other n, $F_n = F_{n-1} + F_{n-2}$.

> **Example 1.4.5**
> Determine a formula for the sum $F_0 + F_1 + F_2 + \ldots + F_n$.
>
> **Solution.** If we collect some initial data we see that the sequence of partial sums is $0, 1, 2, 4, 7, 12, 20, \ldots$. We might recognize these numbers as always one less than a Fibonacci number, and guess that
>
> $$F_0 + F_1 + F_2 + \cdot + F_n = F_{n+2} - 1.$$
>
> We could prove this by induction, but for fun, here is a clever observation that finishes the problem: $F_k = F_{k+2} - F_{k+1}$ (just rearrange the recurrence relation). Now replace each term after F_0:
>
> $$F_0 + (F_3 - F_2) + (F_4 - F_3) + (F_5 - F_4) + \cdots + (F_{n+1} - F_n) + (F_{n+2} - F_{n+1}).$$
>
> Almost every term cancels (telescopes) leaving only
>
> $$-F_1 + F_{n+2}$$
>
> which is of course what we are looking for.

Exercise 90
Determine a formula for the sum $F_0 + F_2 + F_4 + \ldots + F_{2n}$.

Exercise 91
Prove each of the following identities.

(a) $F_{n+1}^2 - F_n^2 = F_{n-1}F_{n+2}$.

(b) $F_k^2 = F_k(F_{k+1} - F_{k-1})$.

Exercise 92
Determine a closed expression for $F_0F_3 + F_1F_4 + \cdots + F_{n-1}F_{n+2}$ using Task 91.a.

Exercise 93
Determine a formula for $F_1^2 + F_2^2 + \ldots + F_n^2$ in two ways. First collect data and then prove your conjecture by mathematical induction. Second, use Task 91.b and telescoping sums.

Exercise 94
Give a geometrical "proof" for the result in Exercise 93.

Exercise 95
Consider the 2×2 matrix $Q = \begin{pmatrix} 0 & 1 \\ 1 & 1 \end{pmatrix}$. Using standard matrix multiplication, compute $Q^2 = Q \cdot Q$, Q^3, and so on. Conjecture a formula for Q^n and prove you are correct. [Hint]

Exercise 96
Prove that $F_{n-1}F_{n+1} - F_n^2 = (-1)^n$ in two ways: First use mathematical induction. Then use Exercise 95 and determinants. There is also a nice "picture proof."

Exercise 97
Is there a result analogous to that in Exercise 96 for just the positive integers $1, 2, 3, 4, \ldots$?

Exercise 98
Show that $Q^{2n+1} = Q^n Q^{n+1}$ and that $F_{2n+1} = F_{n+1}^2 + F_n^2$.

You were asked above to prove the following **Binet formula**:

$$F_n = \frac{a^n - b^n}{a - b}$$

where a and b are the roots of $x^2 - x - 1$. The next example shows an interesting use of this formula.

Example 1.4.6
Prove $\binom{n}{0}F_0 + \binom{n}{1}F_1 + \ldots + \binom{n}{n}F_n = F_{2n}$ in two ways. First, use the Binet Formula from Exercise 87.

Solution. Starting on the right side, use the Binet formula and the fact that a and b satisfy $x^2 - x - 1 = 0$:

$$F_{2n} = \frac{a^{2n} - b^{2n}}{a - b} = \frac{(a+1)^n - (b+1)^n}{a - b}.$$

Then expand the two binomials using the binomial theorem:

$$F_{2n} = \frac{\left[\binom{n}{0} + \binom{n}{1}a + \cdots + \binom{n}{n}a^n\right] - \left[\binom{n}{0} + \binom{n}{1}b + \cdots + \binom{n}{n}b^n\right]}{a - b}.$$

Now group like terms:

$$F_{2n} = \binom{n}{0}\frac{a^0 - b^0}{a - b} + \binom{n}{1}\frac{a - b}{a - b} + \binom{n}{2}\frac{a^2 - b^2}{a - b} + \cdots + \binom{n}{n}\frac{a^n - b^n}{a - b}$$

$$= \binom{n}{0}F_0 + \binom{n}{1}F_1 + \cdots + \binom{n}{n}F_n.$$

You might also try proving this using the matrix Q.

Exercise 99
Prove that $\sum_{n=2}^{\infty} \frac{1}{F_{n-1}F_{n+1}} = 1$. [Hint]

Exercise 100
Prove that $\sum_{n=2}^{\infty} \frac{F_n}{F_{n-1}F_{n+1}} = 2$.

Exercise 101
Prove, directly from the recursion for F_n, that $\lim_{n\to\infty} \frac{F_{n+1}}{F_n} = \frac{1+\sqrt{5}}{2}$.

Exercise 102
For which values of n is F^n a multiple of 3? [Hint]

Exercise 103

Can you have four distinct positive Fibonacci numbers in arithmetic progression?

Exercise 104
Find a closed formula for $\frac{1}{F_1} + \frac{1}{F_2} + \frac{1}{F_4} + \frac{1}{F_8} + \ldots$.

Exercise 105
Conjecture and prove a formula for $\binom{n}{0} + \binom{n-1}{1} + \binom{n-2}{2} + \cdots$.

Exercise 106
Prove that $F_{5n+5} = 3F_{5n} + 5F_{5n+1}$. [Hint]

Exercise 107
Let φ be the positive root of $x^2 - x - 1 = 0$. Prove that $\varphi^n = F_n \varphi + F_{n-1}$.

Exercise 108
Can every positive integer be written as a sum of distinct Fibonacci numbers? In fact, can every positive integer be written as a sum of non-consecutive Fibonacci numbers? Further, can every positive integer be written as a sum of at most 5 non-consecutive Fibonacci numbers? Prove your answers (i.e., either give a proof or provide a counterexample).

1.5 The Catalan Numbers

Let's put all of the techniques we have developed in this chapter to work to solve some deeper counting problems.

1.5.1 A Few Counting Problems

The counting problems in Activities 109–114 are not supposed to be easy to solve, but you have a lot of combinatorial tools to try. Say as much as you can about as many as you can before moving on. At the very least, you should answer the counting problems for small values of n by writing out the set of outcomes. In Subsection 1.5.2, we will investigate the problems in more depth and see how to answer them fully.

Activity 109

We have already seen how to count lattice paths in Activities 2 and 23. Now consider a variation.

How many paths of length $2n$, consisting of horizontal and vertical segments of unit length, are there from $(0,0)$ to (n,n) such that the path never goes above the line $y = x$? One such path to $(3,3)$ is shown in Figure 1.5.1.

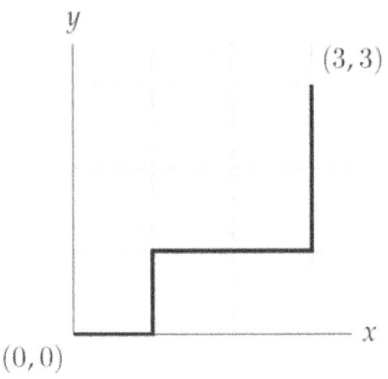

Figure 1.5.1 One of the acceptable lattice paths from $(0,0)$ to $(3,3)$.

Activity 110

$2n$ people stand in line at a old-timey movie theater. Admission is 50 cents, (denoted by H), and the box office starts with no change. n of the people have H and n have $1 (denoted D). In how many ways can the $2n$ people line up so that all can be admitted? [Hint]

Activity 111

Insert the integers $1, 2, \ldots, 2n$ into a 2 by n rectangle of boxes such that the entries are monotonic in rows and columns. We will call this an **acceptable tableau insertion**. How many ways can you do this?

For example, when $n = 3$, there are five arrangements:

1	2	3
4	5	6

1	2	4
3	5	6

1	2	5
3	4	6

1	3	4
2	5	6

1	3	5
2	4	6

Activity 112

If we multiply $n + 1$ numbers, say $a_1 a_2 \cdots a_n a_{n+1}$, we really should put in n pairs of parentheses since multiplication is a binary operation (and what if multiplication is not associative?). How many ways can we do this? For example, when $n = 3$, there are 5 ways:

$$(ab)(cd) \quad ((ab)c)d \quad a(b(cd)) \quad (a(bc))d \quad a((bc)d).$$

[Hint]

Activity 113

Consider a convex polygon with $(n + 2)$ labeled vertices. In how many ways can non-intersecting diagonals be inserted so as to decompose the polygon into triangles? For $n = 3$ the five figures are shown below.

[Hint]

Activity 114

Consider **rooted binary trees**. Rooted trees are trees in the graph theory sense, except that one vertex is designated as the root, which puts a natural ordering on the vertices. Vertices adjacent to the root are its **children**, and vertices adjacent to those (other than the root) are their children, and so on. We are looking at *binary* trees, so each vertex will have either two children (designated the **left child** and **right child**) or no children at all (i.e., the vertex is a leaf).

How many rooted binary trees have exactly $n + 1$ leaves? Note that because we designate children as left or right, the two trees below are counted as distinct.

[Hint]

1.5.2 The Catalan Numbers

The sequence of Catalan numbers, named after Eugene Catalan who along with Euler discovered many of the properties of these numbers, is the sequence $(C_n)_{n \geq 0}$ starting,

$$1, 1, 2, 5, 14, 42, 132, \ldots.$$

All of the counting problems above should be answered by Catalan numbers. Let's investigate this sequence and discover some of its properties.

First, we should make sure that at the very least, all of the counting problems above have the same answer. This will be hugely helpful, since then we could use any of them as a model to discuss the sequence.

How do you prove that two counting problems have the same answer? The bijection principle tells us that if there is a bijection between two sets, then the sets have the same size. So let's find some bijections.

Example 1.5.2

Find a bijection between the set of lattice paths from $(0,0)$ to (n,n) that do not rise above the line $y = x$ (as in Activity 109) and the set of ways that the movie patrons in Activity 110 can line up.

Solution. Represent each lattice path as a string of R's and U's. Represent each way that the movie patrons can line up as a string of H's and D's. The bijection will send each R to an H and each U to a D.

We need to verify that this bijection is both well defined, and that its inverse is well defined (that the function is injective and surjective). The key to all of these is to consider what makes a string valid in each case.

First, notice that the lattice paths must have an equal number of R's and U's. Similarly, the string of movie patrons must have an equal number of H's and D's. Further, every initial substring of R's and U's must have at least as many R's as U's, so that the path does not go above the line $y = x$. Similarly, the every initial substring of H's and D's (half dollars and dollars) must have at least as many H's as D's, so that change can always be given. The H's and R's must *dominate*.

The the conditions on what makes a string valid match up, so every string in one set will match up with exactly one string in the other.

Representing the set of outcomes as strings of sequences is useful, because it suggests a bijection between different types of problems. The strings of paths and movie theater lines from Example 1.5.2 are called **Dyck words**. These are strings of an equal number of two symbols, say x and y, such that no initial segment of the string has more y's than x's.

We will now see that the number of Dyck words of length $2n$ is C_n.

Activity 115

(a) Show that the number of acceptable tableau insertions from Activity 111 is always a Catalan number. That is, give a bijection between the set of Dyck words of length $2n$ to the set of ways to insert the numbers 1 through $2n$ into a $2 \times n$ tableau so that both rows and columns are increasing. [Hint]

(b) Show the number of ways to parenthesize a product of $n+1$ numbers is C_n (see Activity 112). [Hint]

Not all of the models for the Catalan numbers are easily represented by Dyck words (or at least not in an obvious way).

Exercise 116

Demonstrate a bijection between the set of acceptable ways to parenthesize the product of $n + 1$ terms and the set of rooted binary trees with $n + 1$ leaves (see Activity 114). [Hint]

We have now verified that all our sample problems are Catalan numbers, except for the triangulations of polygons problem. There is a clever way to associate each triangulation with a binary tree, as suggested by Figure 1.5.3. However, let's take this opportunity to illustrate another method for proving two problems have the same answers.

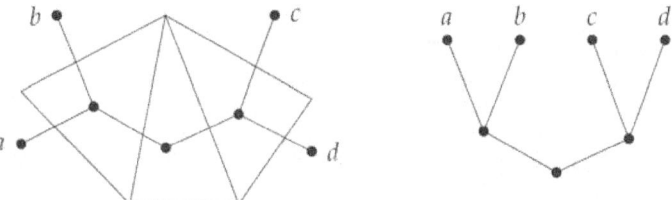

Figure 1.5.3 A triangulation of a 5-gon and associated rooted binary tree.

Activity 117

Consider the recurrence relation

$$C_{n+1} = \sum_{i=0}^{n} C_i C_{n-i} = C_0 C_n + C_1 C_{n-1} + \ldots + C_n C_0$$

with $C_0 = 1$.

(a) Calculate the first 6 values of the sequence from this recurrence relation, to verify that it appears to agree with the Catalan numbers.

(b) Prove that the Catalan numbers satisfy the recurrence relation. [Hint]

(c) Prove that the number of triangulations of a convex polygon with $n + 2$ sides also satisfies this recurrence relation (if you haven't already done so in Exercise 74). Conclude that the triangulations problem is also solved by the Catalan numbers. [Hint]

Next, we would like to find a closed formula for C_n. We can do so using lattice paths.

Activity 118

Recall that C_n gives the number of lattice paths from $(0,0)$ to (n,n) that do not cross the line $y = x$ (but they may touch this line). We will compare this to all the lattice paths from $(0,0)$ to (n,n).

(a) Explain why the number of lattice paths from $(0,0)$ to (n,n) that *do* cross the line $y = x$ is the same as the number of lattice paths from $(0,0)$ to (n,n) that touch or cross the line $y = x + 1$.

(b) Find a bijection between lattice paths from $(0,0)$ to (n,n) that touch (or cross) the line $y = x + 1$ and lattice paths from $(-1, 1)$ to (n, n).
[Hint]

(c) Find a formula for the number of lattice paths from $(0,0)$ to (n,n) that do not cross the line $y = x$. That is, a formula for C_n.
[Hint]

So $C_n = \binom{2n}{n} - \binom{2n}{n+1}$. Another formula you might run into is $C_n = \frac{1}{n+1}\binom{2n}{n}$. Why is this also correct?

Exercise 119

(a) Prove that $\binom{2n}{n} - \binom{2n}{n+1} = \frac{1}{n+1}\binom{2n}{n}$ using a method of your choice.
[Hint]

(b) Challenge problem: explain the formula $C_n = \frac{1}{n+1}\binom{2n}{n+1}$ using the quotient principle. This would give an alternate proof for the closed formula for the Catalan numbers.

1.5.3 More Problems

Here are a few more activities to practice with the Catalan numbers.

Exercise 120

Find C_6 and C_7 using both the recursive and closed formulas.

Exercise 121

Show, by example, how the bijection between Dyck words and valid parenthesizing works. To do this, list the $C_4 = 14$ valid Dyck words, then list the $C_4 = 14$ ways to parenthesize $abcde$, in the corresponding order.

Exercise 122

Here is a way to establish the closed formula for C_n using Dyck words. For simplicity, consider Dyck words using the symbols 0 and 1, insisting that no initial segment contains more 1's than 0's.

(a) Of all $2n$-bit strings of weight n, some are Dyck words and some are not. Explain which are not. What do all of these have in common?

(b) Suppose some initial segment of a bit string contains one more 1 than 0. If you changed every 1 to a 0 and every 0 to a 1 in this initial segment, how many 0's and how many 1's would the entire bit string have? The resulting string will have one more 0 and one fewer 1. That is, if the weight was originally n, the weight would now be $n - 1$.

(c) Describe a bijection between the set of $2n$-bit strings of weight $n - 1$ and the set of $2n$-bit strings of weight n that are NOT Dyck words.

(d) Why does this prove that $C_n = \binom{2n}{n} - \binom{2n}{n-1}$?

Exercise 123
Find a bijection between the ways of parenthesizing $n + 1$ terms and the ways of triangulating a convex polygon with $n + 2$ sides. Illustrate this by matching up the outcomes for $n = 3$.

Exercise 124
Let $C_n = \frac{1}{n+1}\binom{2n}{n}$. Verify the formula:

$$C_n = \binom{n}{1}C_{n-1} - \binom{n-1}{2}C_{n-2} + \binom{n-2}{3}C_{n-3} - \cdots$$

We have considered 6 or 7 different interpretations of Catalan numbers so far. There are many others. In fact, Richard P. Stanley includes 66 different interpretations of the Catalan numbers in the book *Enumerative Combinatorics: Volume 2*.

The remainder of this section contains some examples. See if you can explain why each of these really are interpretations of C_n.

Exercise 125
A and B each receive n votes. How many ways are there that the $2n$ votes can be tallied so that A never trails B.

Exercise 126
Place $2n$ points on the circumference of a circle and draw n non-intersecting chords. How many ways can you do this? Equivalently, if $2n$ people stand in a circle, how many ways can everyone shake hands with one other person so that no one's arms cross? Assume these people have arbitrarily long arms.

Exercise 127
$2n$ kids, all of different heights, must line up for a class picture. They will do this in two equal rows, but for variety, they will take a picture from the front and from the right. How many ways can the kids line up so that both pictures don't have any taller kid blocking a shorter kid?

Exercise 128
Consider the graph P_n (a path with n edges and $n+1$ vertices). Call one of the endpoints of the path v. How many walks of length $2n$ start and stop at v? A walk in a graph is a sequence of adjacent vertices (think of tracing along edges of the graph), that does allow for repeated vertices.

Exercise 129
Draw a n by n right triangle using squares on graph paper (a sort of staircase shape, made up of $\frac{n(n+1)}{2}$ squares). Shade the squares using exactly n colors so that each color makes a rectangle (but not all colors are rectangles of the same dimension). How many ways can this be done?

Exercise 130
If you have $2n$ toothpicks, you could arrange them into a "mountain range" shape by putting half of the toothpicks angled upward, and the other half angled downward. Assume the first toothpick starts at sea level, so no valleys dip below that height. In how many ways can this be done?

Alternatively, we can define a **diagonal lattice path** as one in which each segment travels from (a, b) to $(a+1, b+1)$ or to $(a+1, b-1)$. How many diagonal lattice paths from $(0, 0)$ to $(2n, 0)$ never dip below the x-axis? Such diagonal lattice paths are sometimes called **Dyck paths** or **Catalan paths**.

Exercise 131
How many permutations of $[n]$ do not contain three consecutive numbers abc with $a < b < c$? For example, the permutations 1324 and 1423 are both acceptable, but 2341 and 1342 are not.

Chapter 2

Advanced Combinatorics

The combinatorics we have investigated so far has been *nice*. While not always immediately obvious, each problem had an answer in a nice form that, once you saw how to think about it the right way, could be expressed with a closed formula.

Not all of combinatorics is like this. There are natural questions for which we do not have neat closed formulas. This might be because we (mathematicians) have not yet figured out how to think about these problems in the "right" way, or because there is really no clean approach to solve these problems.

But lack of clean solutions does not mean the problems have no answer. Rather, we must simply look for more abstract techniques before we can get any hold on them.

The plan for this chapter is as follows. We will start with two combinatorial tools. First, the Principle of Inclusion and Exclusion and then generating functions. We will then consider some natural extensions of distribution problems first introduced in Section 1.3: partitioning sets and integers.

2.1 The Principle of Inclusion and Exclusion: the Size of a Union

One of our very first counting principles was the **sum principle** which says that the size of a union of disjoint sets is the sum of their sizes. Computing the size of overlapping sets requires, quite naturally, information about how they overlap. Taking such information into account will allow us to develop a powerful extension of the sum principle known as the "principle of inclusion and exclusion."

2.1.1 Unions of two or three sets

Activity 132
In a biology lab study of the effects of basic fertilizer ingredients on plants, 16 plants are treated with potash, 16 plants are treated with phosphate, and among these plants, eight are treated with both phosphate and potash. No other treatments are used. How many plants receive at least one treatment? If 32 plants are studied, how many receive no treatment?

Activity 133
Give a formula for the size of the union $A \cup B$ of two sets A in terms of the sizes $|A|$ of A, $|B|$ of B, and $|A \cap B|$ of $A \cap B$. If A and B are subsets of

some "universal" set U, express the size of the complement $U - (A \cup B)$ in terms of the sizes $|U|$ of U, $|A|$ of A, $|B|$ of B, and $|A \cap B|$ of $A \cap B$. [Hint]

Activity 134
In Activity 132, there were just two fertilizers used to treat the sample plants. Now suppose there are three fertilizer treatments, and 15 plants are treated with nitrates, 16 with potash, 16 with phosphate, 7 with nitrate and potash, 9 with nitrate and phosphate, 8 with potash and phosphate and 4 with all three. Now how many plants have been treated? If 32 plants were studied, how many received no treatment at all?

Activity 135
A formula for the size of $A \cup B \cup C$ in terms of the sizes of A, B, C and the intersections of these sets is

$$|A \cup B \cup C| = |A| + |B| + |C| - |A \cap B| - |A \cap C| - |B \cap C| + |A \cap B \cap C|.$$

Explain why this is correct using a Venn diagram. In particular, how many times are elements are in $A \cap B \cap C$ counted? What about the other regions of the Venn diagram?

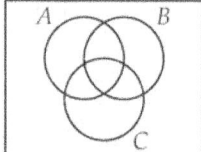

2.1.2 Unions of an arbitrary number of sets

Activity 136
Conjecture a formula for the size of a union of sets

$$A_1 \cup A_2 \cup \cdots \cup A_n = \bigcup_{i=1}^{n} A_i$$

in terms of the sizes of the sets A_i and their intersections.

The difficulty of generalizing Activity 135 and Activity 136 is not likely to be one of being able to see what the right conjecture is, but of finding a good notation to express your conjecture. In fact, it would be easier for some people to express the conjecture in words than to express it in a notation. Here is some notation that will make your task easier. Let us define

$$\bigcap_{i:i \in I} A_i$$

to mean the intersection over all elements i in the set I of A_i. Thus

$$\bigcap_{i:i \in \{1,3,4,6\}} = A_1 \cap A_3 \cap A_4 \cap A_6. \tag{2.1}$$

This kind of notation, consisting of an operator with a description underneath of the values of a dummy variable of interest to us, can be extended in many ways. For example

$$\sum_{I: I \subseteq \{1,2,3,4\}, |I|=2} |\cap_{i \in I} A_i| = |A_1 \cap A_2| + |A_1 \cap A_3| + |A_1 \cap A_4|$$
$$+ |A_2 \cap A_3| + |A_2 \cap A_4| + |A_3 \cap A_4|. \quad (2.2)$$

Exercise 137
Use notation something like that of Equation (2.1) and Equation (2.2) to express the answer to Activity 136. Note there are many different correct ways to do this problem. Try to write down more than one and choose the nicest one you can. Say why you chose it (because your view of what makes a formula nice may be different from somebody else's). The nicest formula won't necessarily involve all the elements of Equations (2.1) and (2.2).

Whether we try to write down a nice expression for the size of a union or whether we just have the basic idea, we should now be able to answer counting questions like the following. We will return to this sort of problem in more detail later, but the adventurous reader can attempt it here.

Exercise 138
A group of n students goes to a restaurant carrying backpacks. The manager invites everyone to check their backpack at the check desk and everyone does. While they are eating, a child playing in the check room randomly moves around the claim check stubs on the backpacks. We will try to compute the probability that, at the end of the meal, at least one student receives his or her own backpack. This probability is the fraction of the total number of ways to return the backpacks in which at least one student gets his or her own backpack back.

(a) What is the total number of ways to pass back the backpacks?

(b) In how many of the distributions of backpacks to students does at least one student get his or her own backpack? [Hint]

(c) What is the probability that at least one student gets the correct backpack?

(d) What is the probability that no student gets his or her own backpack?

(e) As the number of students becomes large, what does the probability that no student gets the correct backpack approach?

Exercise 138 is "classically" called the **hatcheck problem**; the name comes from substituting hats for backpacks. If is also sometimes called the **derangement problem**. A **derangement** of an n-element set is a permutation of that set (thought of as a bijection) that maps no element of the set to itself. One can think of a way of handing back the backpacks as a permutation f of the students: $f(i)$

is the owner of the backpack that student i receives. Then a derangement is a way to pass back the backpacks so that no student gets his or her own.

2.1.3 The Principle of Inclusion and Exclusion

The formula you have given in Exercise 137 is often called **the principle of inclusion and exclusion** for unions of sets. The reason is the pattern in which the formula first adds (includes) all the sizes of the sets, then subtracts (excludes) all the sizes of the intersections of two sets, then adds (includes) all the sizes of the intersections of three sets, and so on. Notice that we haven't yet proved the principle. There are a variety of proofs. Perhaps one of the most straightforward (though not the most elegant) is an inductive proof that relies on the fact that

$$A_1 \cup A_2 \cup \cdots \cup A_n = (A_1 \cup A_2 \cup \cdots \cup A_{n-1}) \cup A_n$$

and the formula for the size of a union of two sets: $|A \cup B| = |A| + |B| - |A \cap B|$.

> **Exercise 139**
> Give a proof of your formula for the principle of inclusion and exclusion.
> [Hint]

Frequently when we apply the principle of inclusion and exclusion, we will have a situation like that of part (d) of Task 138.d. That is, we will have a set A and subsets $A_1, A_2, \ldots, A_n$ and we will want the size or the probability of the set of elements in A that are *not* in the union. This set is known as the **complement** of the union of the A_is in A, and is denoted by $A \setminus \bigcup_{i=1}^{n} A_i$, or if A is clear from context, by $\overline{\bigcup_{i=1}^{n} A_i}$.

Since all the A_is are subsets of A, one way to write this size is as $|A| - \sum_{S:S\subseteq[n], S\neq\emptyset}(-1)^{|S|-1}|\bigcap_{i:i\in S} A_i|$. Letting $|A| = |\bigcap_{i:i\in\emptyset} A_i|$, we may write $\left|\overline{\bigcup_{i=1}^{n} A_i}\right| = \sum_{S:S\subseteq[n]}(-1)^{|S|}|\bigcap_{i:i\in S} A_i|$.

The principle of inclusion and exclusion generally refers to both this formula and the one for the union. Both formulas are notationally horrific, but the idea is straight forward. To find the size of the union, we add the sizes of each set individually, then subtract each of the sizes of the intersections of pairs of sets, then add back in the sizes of the intersections of triples of sets, subtract 4-tuple intersections, add 5-tuple intersections, and so on until we are done. The compliment can be calculated by subtracting the union from the size of the original set.

2.1.4 Application of Inclusion and Exclusion

Multisets with restricted numbers of elements.

> **Exercise 140**
> In how many ways may we distribute k identical apples to n children so that no child gets more than four apples?
> [Hint]

The Menage Problem.

Exercise 141
A group of n married couples comes to a group discussion session where they all sit around a round table. In how many ways can they sit so that no person is next to his or her spouse? (Note that two people of the same sex can sit next to each other.) [Hint]

Exercise 142
A group of n married couples comes to a group discussion session where they all sit around a round table. In how many ways can they sit so that no person is next to his or her spouse or a person of the same sex? This problem is called the **menage problem**. [Hint]

Counting Derangements.

A **derangement** of n elements $\{1, 2, 3, \ldots, n\}$ is a permutation in which no element is fixed. For example, there are 6 permutations of the three elements $\{1, 2, 3\}$:

$$123 \quad 132 \quad 213 \quad 231 \quad 312 \quad 321.$$

but most of these have one or more elements fixed: 123 has all three elements fixed since all three elements are in their original positions, 132 has the first element fixed (1 is in its original first position), and so on. In fact, the only derangements of three elements are

$$231 \text{ and } 312.$$

If we go up to 4 elements, there are $4! = 24$ permutations. How many of these are derangements? If you list out all 24 permutations and eliminate those which are not derangements, you will be left with just 9 derangements. Let's see how we can get that number using PIE.

Activity 143

(a) Write out all 24 permutations of $[4]$ and circle the 9 that are derangements. [Hint]

(b) Locate the permutations that leave 1 fixed (so 1 is the first number in the permutation). How many are there, and does this make sense? Check there are an equal number that leave 2 fixed, and that leave 3 fixed, and that leave 4 fixed.

(c) Notice that there are permutations that leave both 1 and 2 fixed. Which ones are these, and how many are there? Again, verify that there are just as many that leave any pair of numbers fixed.

(d) Repeat the previous task for triples of elements fixed and for all four elements fixed.

(e) We can use the principle of inclusion and exclusion to find the total number of derangements as follows:

$$d_4 = 4! - \left[\binom{4}{1}3! - \binom{4}{2}2! + \binom{4}{3}1! - \binom{4}{4}0! \right].$$

Explain why this is correct. In particular, for the permutation 1234 (where everything is fixed), where in the expression above is it counted?

Of course we can use a similar formula to count the derangements of any number of elements. However, the more elements we have, the longer the formula gets. Here is another example:

Exercise 144
Five gentlemen attend a party, leaving their hats at the door. At the end of the party, they hastily grab hats on their way out. How many different ways could this happen so that none of the gentlemen leave with their own hat?

Derangments are interesting in their own right. In Section 1.2 you proved a few results using combinatorial proofs. In particular, we established the following, which you might try to prove again here.

Theorem 2.1.1
Let d_n denote the number of derangments of n, with $d_0 = 1$ and $d_1 = 0$. Then $d_n = (n-1)(d_{n-1} + d_{n-2})$.

Theorem 2.1.2
Let d_n denote the number of derangments of n, with $d_0 = 1$ and $d_1 = 0$. Then

$$\binom{n}{0}d_0 + \binom{n}{1}d_1 + \binom{n}{2}d_2 + \cdots + \binom{n}{n}d_n = n!$$

Here is another recursion you can explore.

Exercise 145
Use the principle of inclusion and exclusion to derive the recursion

$$d_n = nd_{n-1} + (-1)^n.$$

The non-recursive formula for derangments is messy due to the use of the principle of inclusion and exclusion. Actually, it can be cleaned up a bit, which leads to a remarkable result.

Exercise 146

(a) We know that $d_3 = 3! - \left(\binom{3}{1}2! - \binom{3}{2}1! + \binom{3}{3}0!\right)$. Use the factorial formula for $\binom{n}{k}$ to simplify this. Factor out a $3!$ leaving terms that are each nice unit fractions.

(b) Generalize the previous part to give a nice clean expression for d_n.

(c) The alternating sum of unit fractions you get might look familiar if you have thought about Taylor series from calculus recently. For large n, what fraction of all permutations are derangements? [Hint]

COUNTING SURJECTIVE FUNCTIONS.

There are six surjective functions $f : [3] \to [2]$, which happens to be $2^3 - 2$. This makes sense because there are eight functions all together, but two are not surjective (do you see why?).

Consider functions $f : [3] \to [3]$. Since each element of the domain has three potential images, there are 3^3 such functions. If we only count the injective (one-to-one) functions, we will find $3! = 6$. Since the size of the domain and codomain are equal, this is also the number of surjective functions (since every injective function is surjective in this case).

However, when the size of the codomain is larger than the size of the domain, there will be no injections and plenty of surjections. To see how to count surjections in general, let's explore the case above where we know the answer already.

Activity 147

Write out all 27 functions $f : [3] \to [3]$. You could use two-line notation here, but to save space, perhaps just write the bottom line. So the function that sends 1 to 1, 2 to 3 and 3 to 3 can be represented as (1,3,3).

(a) Circle the functions you found above that are surjective. Verify that there are just 6.

(b) There are various ways that a function might fail to be surjective. One way is for 1 to not be in the range. How many functions are there like this (identify them)? Does this number make sense? Check this is the same number of functions that do not include 2 in the range.

(c) How many functions do not have 1 or 2 in the range? Does this number make sense as well?

(d) Explain what all this has to do with the expression

$$3^3 - \left[\binom{3}{1}2^3 - \binom{3}{2}1^3 + \binom{3}{3}0^3\right].$$

Let's generalize!

Activity 148

Given a function f from the k-element set K to the n-element set $[n]$, we say f is in the set A_i if $f(x) \neq i$ for any x in K.

How many of these sets does a surjective function belong to?

Use this to find the number of functions from a k-element set onto an n-element set?

Exercise 149

If we roll a die eight times, we get a sequence of 8 numbers, the number of dots on top on the first roll, the number on the second roll, and so on.

(a) What is the number of ways of rolling the die eight times so that each of the numbers one through six appears at least once in our sequence? To get a numerical answer, you will likely need a computer algebra package.

(b) What is the probability that we get a sequence in which all six numbers between one and six appear? To get a numerical answer, you will likely need a computer algebra package, programmable calculator, or spreadsheet.

(c) How many times do we have to roll the die to have probability at least one half that all six numbers appear in our sequence? To answer this question, you will likely need a computer algebra package, programmable calculator, or spreadsheet.

2.2 Generating Functions

In this section we will consider how to use polynomials to help us solve counting problems. Why would we even think there would be a connection here?

Activity 150

Suppose you have two number cubes with non-standard labels. The first has 1's on four sides and 2's on the remaining two sides, while the second cube has two 1's and four 2's. You will roll the pair of dice and compute the sum of the numbers rolled.

(a) How many ways could you roll a 2? What about a 3? 4? You should be able to answer these questions by making a 6 × 6 addition table (with a lot of repeated row and column headers), but make sure you also think about how the sum and product principles are being used.

(b) Expand the polynomial $(4x + 2x^2)(2x + 4x^2)$. Do this by hand, keeping track of what you get before and after you "group like terms."

(c) How are the two tasks above related? What is going on here?

Activity 151

A standard problem to use when introducing probability is to ask for the probability of rolling a particular sum on two regular number cubes (dice numbered 1 through 6). Students might not immediately see that some sums are more likely to occur than others. You can illustrate this by creating a table.

(a) Complete the standard addition table used to count the number of ways each sum of two regular dice can be achieved. What would you do with the entries in the table to find the number of ways to roll an 8, for example?

(b) How is the table above related to what you would get if you complete the table below?

	x	x^2	x^3	x^4	x^5	x^6
x						
x^2						
x^3						
x^4						
x^5						
x^6						

In particular, what should the operation on the table be to make the correspondence relevant? [Hint]

(c) What do each of the tasks above have to do with expanding $(x + x^2 + \cdots + x^6)^2$

Exercise 152

Using the connection between dice and polynomials, we can find interesting ways to label dice without modifying probabilities.

(a) Show algebraically that $x+x^2+x^3+x^4+x^5+x^6 = x(1+x+x^2)(1+x)(1-x+x^2)$. Try to do this by factoring instead of multiplying out the right-hand side.

(b) Multiply out the polynomial $x(x+1)(x^2+x+1)$ and also $x(x+1)(x^2+x+1)(x^2-x+1)^2$.

(c) Explain how the previous two parts together allow you to find another way to label dice so that the probability of throwing any number 2 to 12 is the same as for standard dice. [Hint]

Activity 153

It is also possible to multiply infinite polynomials (i.e., power series). As long as we are only interested in the first few terms, we can find these by multiplying a small number of terms from the two infinite polynomials. Multiply the following series out through at least x^7.

(a) Multiply $(1+x+x^2+x^3+\cdots)(1+x+x^2+x^3+\cdots)$.

(b) Multiply $(1+x+x^2+x^3+\cdots)(x+2x^2+3x^3+4x^4+\cdots)$.

(c) What is going on here? Where have we seen the coefficients of these polynomials before?

2.2.1 Visualizing Counting with Pictures

Suppose you are going to choose three pieces of fruit from among apples, pears and bananas for a snack. We can symbolically represent all your choices as

🍎🍎🍎 + 🍐🍐🍐 + 🍌🍌🍌 + 🍎🍎🍐 + 🍎🍎🍌 + 🍎🍐🍐 + 🍐🍐🍌 + 🍎🍌🍌 + 🍐🍌🍌 + 🍎🍐🍌.

Here we are using a picture of a piece of fruit to stand for taking a piece of that fruit. Thus 🍎 stands for taking an apple, 🍎🍐 for taking an apple and a pear, and 🍎🍎 for taking two apples. You can think of the plus sign as standing for the "exclusive or," that is, 🍎 + 🍌 would stand for "I take an apple or a banana but not both." To say "I take both an apple and a banana," we would write 🍎🍌. We can extend the analogy to mathematical notation by condensing our statement that we take three pieces of fruit to

$🍎^3 + 🍐^3 + 🍌^3 + 🍎^2🍐 + 🍎^2🍌 + 🍎🍐^2 + 🍐^2🍌 + 🍎🍌^2 + 🍐🍌^2 + 🍎🍐🍌.$

In this notation $🍎^3$ stands for taking a multiset of three apples, while $🍎^2🍌$ stands for taking a multiset of two apples and a banana, and so on. What our notation is really doing is giving us a convenient way to list all three element multisets chosen from the set $\{🍎, 🍐, 🍌\}$.[1]

Suppose now that we plan to choose between one and three apples, between one and two pears, and between one and two bananas. In a somewhat clumsy way we could describe our fruit selections as

$$\begin{array}{lll}
🍎🍐🍌 + 🍎^2🍐🍌 & +\cdots+ 🍎^2🍐^2🍌 & +\cdots+ 🍎^2🍐^2🍌^2 \\
+ 🍎^3🍐🍌 & +\cdots+ 🍎^3🍐^2🍌 & +\cdots+ 🍎^3🍐^2🍌^2.
\end{array} \qquad (2.3)$$

Activity 154

Using an A in place of the picture of an apple, a P in place of the picture of a pear, and a B in place of the picture of a banana, write out the formula similar to Formula (2.3) without any dots for left out terms. (You may use pictures instead of letters if you prefer, but it gets tedious quite quickly!) Now expand the product $(A + A^2 + A^3)(P + P^2)(B + B^2)$ and compare the result with your formula.

Activity 155

Substitute x for all of A, P and B (or for the corresponding pictures) in the formula you got in Activity 154 and expand the result in powers of x. Give an interpretation of the coefficient of x^n.

If we were to expand the formula

$$(🍎 + 🍎^2 + 🍎^3)(🍐 + 🍐^2)(🍌 + 🍌^2). \qquad (2.4)$$

we would get Formula (2.3). Thus Formula (2.3) and Formula (2.4) each describe the number of multisets we can choose from the set $\{🍎, 🍐, 🍌\}$ in which 🍎 appears between 1 and three times and 🍐 and 🍌 each appear once or twice. We interpret Formula (2.3) as describing each individual multiset we can choose, and we interpret Formula (2.4) as saying that we first decide how many apples to take, and then decide how many pears to take, and then decide how many bananas to take. At this stage it might seem a bit magical that doing ordinary algebra with the second formula yields the first, but in fact we could define addition and multiplication with these pictures more formally so we could explain in detail why things work out. However since the pictures are for motivation, and are actually difficult to write out on paper, it doesn't make much sense to work out these details. We will see an explanation in another context later on.

2.2.2 Picture functions

As you've seen, in our descriptions of ways of choosing fruits, we've treated the pictures of the fruit as if they are variables. You've also likely noticed that it is much easier to do algebraic manipulations with letters rather than pictures, simply because it is time consuming to draw the same picture over and over again,

---

[1] This approach was inspired by George Pólya's paper "Picture Writing," in the December, 1956 issue of the *American Mathematical Monthly*, page 689. While we are taking a somewhat more formal approach than Pólya, it is still completely in the spirit of his work.

while we are used to writing letters quickly. In the theory of generating functions, we associate variables or polynomials or even power series with members of a set. There is no standard language describing how we associate variables with members of a set, so we shall invent[2] some. By a **picture** of a member of a set we will mean a variable, or perhaps a product of powers of variables (or even a sum of products of powers of variables). A function that assigns a picture $P(s)$ to each member s of a set S will be called a **picture function**. The **picture enumerator** for a picture function P defined on a set S will be

$$E_P(S) = \sum_{s : s \in S} P(s).$$

We choose this language because the picture enumerator lists, or enumerates, all the elements of S according to their pictures. Thus Formula (2.3) is the picture enumerator the set of all multisets of fruit with between one and three apples, one and two pears, and one and two bananas.

Exercise 156
How would you write down a polynomial in the variable A that says you should take between zero and three apples?

Exercise 157
How would you write down a picture enumerator that says we take between zero and three apples, between zero and three pears, and between zero and three bananas?

Notice that when we used A^2 to stand for taking two apples, and P^3 to stand for taking three pears, then we used the product $A^2 P^3$ to stand for taking two apples and three pears. Thus we have chosen the picture of the ordered pair (2 apples, 3 pears) to be the product of the pictures of a multiset of two apples and a multiset of three pears.

Exercise 158
Show that if S_1 and S_2 are sets with picture functions P_1 and P_2 defined on them, and if we define the picture of an ordered pair $(x_1, x_2) \in S_1 \times S_2$ to be $P((x_1, x_2)) = P_1(x_1) P_2(x_2)$, then the picture enumerator of P on the set $S_1 \times S_2$ is $E_{P_1}(S_1) E_{P_2}(S_2)$. We call this the **product principle for picture enumerators**.

2.2.3 Generating functions

Example 2.2.1
Suppose you are going to choose a snack of between zero and three apples, between zero and three pears, and between zero and three bananas. Write

---
[2] We are really adapting language introduced by George Pólya.

down a polynomial in one variable x such that the coefficient of x^n is the number of ways to choose a snack with n pieces of fruit.

Solution. We have a polynomial in three variables A, P and B for this situation:

$$(1 + A + A^2 + A^3)(1 + P + P^2 + P^3)(1 + B + B^2 + B^3).$$

Now we substitute x for each of those variables to get

$$(1 + x + x^2 + x^3)^3$$

Exercise 159
Suppose an apple costs 20 cents, a banana costs 25 cents, and a pear costs 30 cents. What should you substitute for A, P, and B in Exercise 157 in order to get a polynomial in which the coefficient of x^n is the number of ways to choose a selection of fruit that costs n cents? [Hint]

Exercise 160
Suppose an apple has 40 calories, a pear has 60 calories, and a banana has 80 calories. What should you substitute for A, P, and B in Exercise 157 in order to get a polynomial in which the coefficient of x^n is the number of ways to choose a selection of fruit with a total of n calories?

Exercise 161
We are going to choose a subset of the set $\{1, 2, \ldots, n\}$. Suppose we use x_1 to be the picture of choosing 1 to be in our subset. What is the picture enumerator for either choosing 1 or not choosing 1? Suppose that for each i between 1 and n, we use x_i to be the picture of choosing i to be in our subset. What is the picture enumerator for either choosing i or not choosing i to be in our subset? What is the picture enumerator for all possible choices of subsets of $[n]$? What should we substitute for x_i in order to get a polynomial in x such that the coefficient of x^k is the number of ways to choose a k-element subset of n? What theorem have we just reproved (a special case of)? [Hint]

In Exercise 161 we see that we can think of the process of expanding the polynomial $(1 + x)^n$ as a way of "generating" the binomial coefficients $\binom{n}{k}$ as the coefficients of x^k in the expansion of $(1 + x)^n$. For this reason, we say that $(1 + x)^n$ is the "generating function" for the binomial coefficients $\binom{n}{k}$. More generally, the **generating function** for a sequence a_i, defined for i with $0 \le i \le n$ is the expression $\sum_{i=0}^{n} a_i x^i = a_0 + a_1 x + a_2 x^2 + \cdots + a_n x^n$, and the **generating function** for the sequence a_i with $i \ge 0$ is the expression $\sum_{i=0}^{\infty} a_i x^i$.

Example 2.2.2
The generating function for the sequence $1, 2, 4, 8, \ldots$ is
$$1 + 2x + 4x^2 + 8x^3 + \cdots$$
which we will see later can be expressed as
$$1 + (2x) + (2x)^2 + (2x)^3 + \cdots = \frac{1}{1 - 2x}.$$
The sequence $1, 3, 5, 7, \ldots$ has generating function
$$1 + 3x + 5x^2 + 7x^3 + \cdots = \frac{1+x}{(1-x)^2}.$$
We will soon see why you can write the infinite series in this nice way.

Generating functions for infinite sequences are examples of power series. In calculus it is important to think about whether a power series converges in order to determine whether or not it represents a function. In a nice twist of language, even though we use the phrase generating function as the name of a power series in combinatorics, we don't require the power series to actually represent a function in the usual sense, and so we don't have to worry about convergence.[3] Instead we think of a power series as a convenient way of representing the terms of a sequence of numbers of interest to us. The only justification for saying that such a representation is convenient is because of the way algebraic properties of power series capture some of the important properties of some sequences that are of combinatorial importance. The remainder of this chapter is devoted to giving examples of how the algebra of power series reflects combinatorial ideas.

Because we choose to think of power series as strings of symbols that we manipulate by using the ordinary rules of algebra and we choose to ignore issues of convergence, we have to avoid manipulating power series in a way that would require us to add infinitely many real numbers. For example, we cannot make the substitution of $y + 1$ for x in the power series $\sum_{i=0}^{\infty} x^i$, because in order to interpret $\sum_{i=0}^{\infty} (y+1)^i$ as a power series we would have to apply the binomial theorem to each of the $(y+1)^i$ terms, and then collect like terms, giving us infinitely many ones added together as the coefficient of y^0, and in fact infinitely many numbers added together for the coefficient of any y^i. (On the other hand, it would be fine to substitute $y + y^2$ for x. Can you see why?)

2.2.4 Power series

For now, most of our uses of power series will involve just simple algebra. Since we use power series in a different way in combinatorics than we do in calculus, we should review a bit of the algebra of power series.

---

[3]In the evolution of our current mathematical terminology, the word function evolved through several meanings, starting with very imprecise meanings and ending with our current rather precise meaning. The terminology "generating function" may be thought of as an example of one of the earlier usages of the term function.

Activity 162

Multiply out the polynomial $(a_0 + a_1x + a_2x^2)(b_0 + b_1x + b_2x^2 + b_3x^3)$. What is the coefficient of x^2? What is the coefficient of x^4?

Activity 163

In Activity 162 why is there a b_0 and a b_1 in your expression for the coefficient of x^2 but there is not a b_0 or a b_1 in your expression for the coefficient of x^4? What is the coefficient of x^4 in

$$(a_0 + a_1x + a_2x^2 + a_3x^3 + a_4x^4)(b_0 + b_1x + b_2x^2 + b_3x^3 + b_4x^4)?$$

Express this coefficient in the form

$$\sum_{i=0}^{4} \text{something},$$

where the something is an expression you need to figure out. Now suppose that $a_3 = 0, a_4 = 0$ and $b_4 = 0$. To what is your expression equal after you substitute these values? In particular, what does this have to do with Activity 162? [Hint]

Activity 164

The point of the Problems 162 and Activity 163 is that so long as we are willing to assume $a_i = 0$ for $i > n$ and $b_j = 0$ for $j > m$, then there is a very nice formula for the coefficient of x^k in the product

$$\left(\sum_{i=0}^{n} a_i x^i\right)\left(\sum_{j=0}^{m} b_j x^j\right).$$

Write down this formula explicitly. [Hint]

Activity 165

Assuming that the rules you use to do arithmetic with polynomials apply to power series, write down a formula for the coefficient of x^k in the product

$$\left(\sum_{i=0}^{\infty} a_i x^i\right)\left(\sum_{j=0}^{\infty} b_j x^j\right).$$

[Hint]

We use the expression you obtained in Activity 165 to *define* the product of power series. That is, we define the product

$$\left(\sum_{i=0}^{\infty} a_i x^i\right)\left(\sum_{j=0}^{\infty} b_j x^j\right)$$

to be the power series $\sum_{k=0}^{\infty} c_k x^k$, where c_k is the expression you found in Activity 165. Since you derived this expression by using the usual rules of algebra for polynomials, it should not be surprising that the product of power series satisfies these rules.[4]

2.2.5 The extended binomial theorem and multisets

The binomial theorem tells us how to expand the binomial $(1+x)^n$:

$$(1+x)^n = \binom{n}{0} + \binom{n}{1}x + \binom{n}{2}x^2 + \cdots + \binom{n}{n}x^n.$$

This works for any positive integer n.

We will now consider the *extended* binomial theorem, which tells us what to do if n is a negative integer. In particular, we will see how to expand $(1+x)^{-n} = \frac{1}{(1+x)^n}$ (here taking n positive again).

Although we could use calculus-like tricks to manipulate series directly (and we will demonstrate this in a bit), let's first try to develop the formula combinatorially using generating functions. After all, $(1+x)^n$ is the generating function for the number k-element subsets of $[n]$ (you should be able to say why for two good reasons now).

What happens when we instead count the number of k-element *multisets* of $[n]$? Remember, we saw in Theorem 1.3.3 that this can be expressed as the binomial coefficeint $\left(\!\binom{n}{k}\!\right) = \binom{n+k-1}{k}$. Let's put this into a generating function.

Activity 166

Suppose that i is an integer between 1 and n.

(a) What is the generating function in which the coefficient of x^k is 1? This series is an example of what is called an **infinite geometric series**.

(b) Express the generating function in which the coefficient of x^k is the number of k-element multisets chosen from $[n]$ as a *power* of a power series.

Note, it is useful to interpret the coefficient 1 as the number of multisets of size k chosen from the *singleton* set $\{i\}$. Namely, there is only one way to choose a multiset of size k from $\{i\}$: choose i exactly k times.

What does this, together with the binomial coefficient for multisets tell you about what this generating function equals? [Hint]

Activity 167

Compute the product $(1-x)\sum_{k=0}^{n} x^k = (1-x)(1+x+x^2+\cdots+x^n)$. Then compute the product $(1-x)\sum_{k=0}^{\infty} x^k$.

---

[4]Technically we should explicitly state these rules and prove that they are all valid for power series multiplication, but it seems like overkill at this point to do so!

Another way to arrive at the conclusion of the previous activity is the following. Set $S = 1 + x + x^2 + \cdots$. Then $xS = x + x^2 + x^3 + \cdots$. Consider the difference $S - xS$. Combining like terms we find $S - xS = (1-x)S = 1$, as every power of x cancels except the first. Therefore we find

$$1 + x + x^2 + \cdots = \frac{1}{1-x}.$$

Activity 168
Express the generating function for the number of multisets of size k chosen from $[n]$ (where n is fixed but k can be any nonnegative integer) as a 1 over something relatively simple.

We can now put this all together to get a special case of the extended binomial theorem.

Activity 169
Find a formula for $(1+x)^{-n} = \frac{1}{(1+x)^n}$ as a power series whose coefficients involve binomial coefficients. [Hint]

For variety, here is another way you can get to the extended binomial theorem.

Activity 170
We start with the geometric series

$$\frac{1}{1-x} = 1 + x + x^2 + x^3 + \cdots.$$

(a) What happens if we take a derivative of both sides? Do this to get a power series for $\frac{1}{(1-x)^2}$. [Hint]

(b) Take another derivative and divide both sides by 2 to get a power series for $\frac{1}{(1-x)^3}$. Then write the coefficients in the power series as binomial coefficients (use the factorial formula).

(c) Prove, by mathematical induction on n that

$$\frac{1}{(1-x)^n} = \binom{n-1}{0} + \binom{n}{1}x + \binom{n+1}{2}x^2 + \cdots + \binom{n+k-1}{k}x^k + \cdots.$$

(d) Finally, substitute $-x$ for x on both sides.

We used generating functions for multisets to get the extended binomial theorem, but these generating functions are useful in themselves. We close with a few counting problems related to these.

Exercise 171
Write down the generating function for the number of ways to distribute identical pieces of candy to three children so that no child gets more than

4 pieces. Write this generating function as a quotient of polynomials. Using both the extended binomial theorem and the original binomial theorem, find out in how many ways we can pass out exactly ten pieces.

[Hint]

Exercise 172
Another way to count the number of ways to distribute 10 identical pieces of candy to 3 children so no child gets more than 4 is to use the Principle of Inclusion and Exclusion. Do this, and compare this calculation to what you found in Exercise 171.

[Hint]

Exercise 173
What is the generating function for the number of multisets chosen from an n-element set so that each element appears at least j times and less than m times? Write this generating function as a quotient of polynomials, then as a product of a polynomial and a power series.

[Hint]

2.2.6 Generating Functions and Recurrence Relations

Recall that a recurrence relation for a sequence a_n expresses a_n in terms of values a_i for $i < n$. For example, the equation $a_i = 3a_{i-1} + 2^i$ is a first order linear constant coefficient recurrence.

Algebraic manipulations with generating functions can sometimes reveal the solutions to a recurrence relation.

Example 2.2.3
Find a generating functions for the recurrence relation $a_n = 3a_{n-1} + 4a_{n-2}$ with initial conditions $a_1 = 3$ and $a_2 = 13$ (this is the same recurrence relation we solved using the characteristic root technique in Example 1.4.4).

Solution. In general, we have some generating function $f(x) = a_0 + a_1 x + a_2 x^2 + \cdots$. In this case, we know that $a_1 = 3$ and $a_2 = 13$; we can find $a_3 = 51$ and, by working backwards, $a_0 = 1$. Having these values makes writing the following out a little clearer.

In addition to $f(x)$ we will also consider $3x \cdot f(x)$ and $4x^2 f(x)$.

$$f(x) = 1 + 3x + 13x^2 + 51x^3 + \cdots$$
$$3x f(x) = 3x + 9x^2 + 39x^3 + 153x^4 +$$
$$4x^2 f(x) = 4x^2 + 12x^3 + 52x^4 + 204x^5 +$$

Now subtract: $f(x) - 3xf(x) - 4x^2 f(x) = 1$. That's it. Everything else on the right and side cancels. Why? Well, if you consider the x^n term, it will be $a_n x^n - 3a_{n-1} x^n - 4a_{n-2} x^n$. The recurrence relation tells us that the coefficient will be zero!

Simplifying this a bit more, we find $[1 - 3x - 4x^2]f(x) = 1$, or that the generating function for the sequence is

$$f(x) = \frac{1}{1 - 3x - 4x^2}.$$

Now the question becomes, why does this help? We would like to find a closed formula for a_n. Right now, we just know that a_n is the coefficient of x^n in the power series for $\frac{1}{1-3x-4x^2}$. We will come back to this in the next subsection.

Exercise 174
Find the generating function for the solutions to the recurrence

$$a_i = 5a_{i-1} - 6a_{i-2}.$$

Of course there is no need to use generating functions to solve recurrences like the ones above, since we could use the characteristic root technique. It gets more interesting when we consider recurrence relations for which the characteristic root technique fails.

Activity 175
Consider the recurrence relation $a_i = 3a_{i-1} + 3^i$ with initial term $a_0 = 1$.

(a) Write out the first few terms of the sequence. Then write down the generating series $f(x) = a_0 + a_1 x + a_2 x^2 + \cdots$ (that is, just substitute in the right values for a_i).

(b) Write out the series for $3x f(x)$. Does it make sense why $3x$ is the right thing to multiply by?

(c) Now subtract $f(x) - 3x f(x)$ to get a series for $(1 - 3x) \cdot f(x)$. Show how this proves that the generating function for (a_n) is

$$f(x) = \frac{1}{(1 - 3x)^2}.$$

(d) Now explain why the coefficient of x^n is $\binom{n+1}{1}3^n$ and what that tells us about the sequence (a_n).

Exercise 176
Use the previous approach to solve the recurrence relation $a_i = 3a_{i-1} + 2^i$. Do this in general, for any initial value a_0. [Hint]

Exercise 177
Suppose we deposit $5000 in a savings certificate that pays ten percent interest and also participate in a program to add $1000 to the certificate at

the end of each year (from the end of the first year on) that follows (also subject to interest.) Assuming we make the $5000 deposit at the end of year 0, and letting a_i be the amount of money in the account at the end of year i, write a recurrence for the amount of money the certificate is worth at the end of year n. Solve this recurrence. How much money do we have in the account (after our year-end deposit) at the end of ten years? At the end of 20 years?

The sequence of problems that follows (culminating in Exercise 188) describes a number of hypotheses we might make about a fictional population of rabbits. We use the example of a rabbit population for historic reasons; our goal is a classical sequence of numbers called Fibonacci numbers. When Fibonacci[5] introduced them, he did so with a fictional population of rabbits.

Exercise 178

Suppose we start (at the end of month 0) with 10 pairs of baby rabbits, and that after baby rabbits mature for one month they begin to reproduce, with each pair producing two new pairs at the end of each month afterwards. Suppose further that over the time we observe the rabbits, none die.

Let a_n be the number of rabbits we have at the end of month n. So we have $a_0 = 10$, $a_1 = 10$, $a_2 = 30$, and so on. Show that $a_n = a_{n-1} + 2a_{n-2}$. This is an example of a **second order** *linear* recurrence with constant coefficients.

Using a method similar to that of Activity 175, show that

$$\sum_{i=0}^{\infty} a_i x^i = \frac{10}{1 - x - 2x^2}.$$

This gives us the generating function for the sequence a_i giving the population in month i; shortly we shall see a method for converting this to a solution to the recurrence.

Exercise 179

In Fibonacci's original problem, each pair of mature rabbits produces one new pair at the end of each month, but otherwise the situation is the same as in Exercise 178. Assuming that we start with one pair of baby rabbits (at the end of month 0), find the generating function for the number of pairs of rabbits we have at the end of n months. [Hint]

Exercise 180

Use long division on $\dfrac{x}{1 - x - x^2}$ and see what pops out.

---

[5]Apparently Leanardo de Pisa was given the name Fibonacci posthumously. It is a shortening of "son of Bonacci" in Italian.

In the next section we will see how to use the generating functions for these recursions to find closed formulas for the sequence, including a closed formula for the Fibonacci numbers. While we are here though, consider the following example of another reason why having a generating function for the Fibonacci numbers is useful.

Example 2.2.4
Use generating functions to prove the identity $F_{n+1} = \binom{n}{0} + \binom{n-1}{1} + \binom{n-2}{2} + \cdots$ (note this is a finite sum).

Solution. Start with the generating function for the sequence $1, 1, 2, 3, 5,$ which we saw in Exercise 179 is

$$\frac{1}{1-x-x^2} = 1 + x + 2x^2 + \cdots + F_{n+1}x^n + \cdots.$$

Of course we also have

$$\frac{1}{1-x} = 1 + x + x^2 + \cdots$$

and this is almost the same thing. In fact, if we replace x with $x + x^2$ we get the Fibonacci generating function:

$$\frac{1}{1-(x+x^2)} = 1 + (x+x^2) + (x+x^2)^2 + \cdots$$
$$= 1 + x(1+x) + x^2(1+x)^2 + x^3(1+x)^3 + \cdots.$$

Now use the binomial theorem on each of the $(1+x)^n$ terms and distribute:

$$1 + x + x^2 + (x^2 + 2x^3 + x^4)$$
$$+ (x^3 + 3x^4 + 3x^5 + x^6)$$
$$+ (x^4 + 4x^5 + 6x^6 + 4x^7 + x^8) + \cdots.$$

Look at the coefficient of x^4, for example. We have $F_5 = 1 + 3 + 1 = \binom{4}{0} + \binom{3}{1} + \binom{2}{2}$. It should be clear that this works in general: the coefficient of x^n will be F_{n+1} on the one hand, and $\binom{n}{0} + \binom{n-1}{1} + \binom{n-2}{2} + \cdots$ on the other.

We also point out that there are other techniques for getting the generating function for a sequence other than the "differencing" we have used above. We illustrate this with an extreme example.

Example 2.2.5
Find the generating function for the sequence (a_n) where $a_n = \binom{2n}{n}$.

Solution. We have seen that the power series $1 + x + x^2 + \cdots$ can be written as $\frac{1}{1-x}$. Here we wish to find a nice expression for

$$\binom{1}{0} + \binom{2}{1}x + \binom{4}{2}x^2 + \binom{6}{3}x^3 + \cdots.$$

First we note a sort of recurrence: $(n + 1)a_{n+1} = (4n + 2)a_n$. To verify this use the factorial formula for $\binom{n}{k}$:

$$(n+1)a_{n+1} = (n+1)\binom{2n+2}{n+1} = (n+1)\frac{(2n+2)(2n+1)}{(n+1)!(n+1)!}(2n)! = (4n+2)\binom{2n}{n}.$$

Now, $f(x) = a_0 + a_1 x + a_2 x^2 + \cdots = \sum_{n=0}^{\infty} a_n x^n$ has derivative

$$f'(x) = a_1 + 2a_2 x + 3a_3 x^2 + \cdots = \sum_{n=0}^{\infty}(n+1)a_{n+1}x^n$$

and further

$$xf'(x) = a_1 x + 2a_2 x^2 + 3a_3 x^3 + \cdots = \sum_{n=0}^{\infty} n a_n x^n.$$

Using the recurrence above, we have

$$\sum_{n=0}^{\infty}(n+1)a_{n+1}x^n = 4\sum_{n=0}^{\infty} n a_n x^n + 2\sum_{n=0}^{\infty} a_n x^n.$$

This means

$$f'(x) = 4xf'(x) + 2f(x)$$

or equivalently

$$\frac{f'(x)}{f(x)} = \frac{2}{1-4x}.$$

Now integrate! We get

$$\ln f(x) = -\frac{1}{2}\ln(1-4x) + c.$$

Since $a_0 = 1$, it must be $c = 0$. Finally then, by exponentiating, we arrive at

$$f(x) = \frac{1}{\sqrt{1-4x}}.$$

2.2.7 Partial fractions

The generating functions you found in the previous section all can be expressed in terms of the reciprocal of a quadratic polynomial. However without a power series representation, the generating function doesn't tell us what the sequence is. It turns out that whenever you can factor a polynomial into linear factors (and over the complex numbers such a factorization always exists) you can use that factorization to express the reciprocal in terms of power series.

Activity 181
Express $\frac{1}{x-3} + \frac{2}{x-2}$ as a single fraction.

Activity 182
In Activity 181 you see that when we added numerical multiples of the reciprocals of first degree polynomials we got a fraction in which the denominator is a quadratic polynomial. This will always happen unless the two denominators are multiples of each other, because their least common multiple will simply be their product, a quadratic polynomial. This leads us to ask whether a fraction whose denominator is a quadratic polynomial can always be expressed as a sum of fractions whose denominators are first degree polynomials. Find numbers c and d so that

$$\frac{5x+1}{(x-3)(x+5)} = \frac{c}{x-3} + \frac{d}{x+5}.$$

[Hint]

Activity 183
In Activity 182 you may have simply guessed at values of c and d, or you may have solved a system of equations in the two unknowns c and d. Given constants a, b, r_1, and r_2 (with $r_1 \neq r_2$), write down a system of equations we can solve for c and d to write

$$\frac{ax+b}{(x-r_1)(x-r_2)} = \frac{c}{x-r_1} + \frac{d}{x-r_2}.$$

[Hint]

Writing down the equations in Activity 183 and solving them is called the **method of partial fractions**. This method will let you find power series expansions for generating functions of the type you found in Problems 178 to Exercise 174. However you have to be able to factor the quadratic polynomials that are in the denominators of your generating functions.

Example 2.2.6
Let's return to our familiar recurrence relation $a_n = 3a_{n-1} + 4a_{n-2}$ with initial terms $a_1 = 3$ and $a_2 = 13$.
In Example 1.4.4 we used the characteristic root technique to find the closed formula

$$a_n = \frac{4}{5}4^n + \frac{1}{5}(-1)^n.$$

In Example 2.2.3 we found that the generating function for the sequence was
$$f(x) = \frac{1}{1 - 3x - 4x^2}.$$
We can use partial fractions to reconcile these two results.

Note that the denominator factors as $(1+x)(1-4x)$. We write
$$\frac{1}{1-3x-4x^2} = \frac{c}{1+x} + \frac{d}{1-4x}$$
and solve for c and d. This is done by finding a common denominator and equating the numerators to get
$$1 = c(1-4x) + d(1+x).$$
Note that if $x = -1$ this gives use $1 = 5c$ so $c = \frac{1}{5}$. Similarly, if $x = \frac{1}{4}$ we find $1 = \frac{5}{4}d$ so $d = \frac{4}{5}$. (These constants should look familiar.)

So what? The generating function for a_n can now be expressed as the sum of two generating functions: $\frac{1/5}{1+x}$ and $\frac{4/5}{1-4x}$. What's more, these generating functions have nice power series:
$$\frac{1/5}{1+x} = \frac{1}{5}\left(1 - x + x^2 - x^3 + \cdots + (-1)^n x^n + \cdots\right);$$
$$\frac{4/5}{(1-4x)} = \frac{4}{5}\left(1 + 4x + 16x^2 + \cdots + 4^n x^n + \cdots\right).$$
Both of these were found by substituting a value (either $-x$ or $4x$) into the power series for $\frac{1}{1-x}$.

Since a_n is the coefficient of x^n, we get
$$a_n = \frac{1}{5}(-1)^n + \frac{4}{5}4^n$$
as expected.

Exercise 184
Use the method of partial fractions to convert the generating function $\frac{10}{1-x-2x^2}$ from Exercise 178 into the sum of two nicer generating functions. Use this to find a formula for a_n.

Exercise 185
Find the value of a_{40} in the power series
$$\frac{x-2}{3-4x+x^2} = a_0 + a_1 x + a_2 x^2 + \cdots + a_{40} x^{40} + \cdots.$$

[Hint]

Exercise 186
Use the quadratic formula to find the solutions to $x^2 + x - 1 = 0$, and use that information to factor $x^2 + x - 1$.

Exercise 187
Use the factors you found in Exercise 186 to write
$$\frac{1}{x^2 + x - 1}$$
in the form
$$\frac{c}{x - r_1} + \frac{d}{x - r_2}.$$
[Hint]

Exercise 188

(a) Use the partial fractions decomposition you found in Exercise 187 to write the generating function you found in Exercise 179 in the form
$$\sum_{n=0}^{\infty} a_n x^i$$
and use this to give an explicit formula for a_n. [Hint]

(b) When we have $a_0 = 1$ and $a_1 = 1$, i.e. when we start with one pair of baby rabbits, the numbers a_n are called **Fibonacci Numbers**. Use either the recurrence or your final formula to find a_2 through a_8. Are you amazed that your general formula produces integers, or for that matter produces rational numbers? Why does the recurrence equation tell you that the Fibonacci numbers are all integers?

(c) Explain why there is a real number b such that, for large values of n, the value of the nth Fibonacci number is almost exactly (but not quite) some constant times b^n. (Find b and the constant.)

(d) Find an algebraic explanation (not using the recurrence equation) of what happens to make the square roots of 5 go away. [Hint]

(e) As a challenge (which the authors have not yet done), see if you can find a way to show algebraically (not using the recurrence relation, but rather the formula you get by removing the square roots of five) that the formula Binet for the Fibonacci numbers yields integers.

---

[6]We use the words roots and solutions interchangeably.

84   2. ADVANCED COMBINATORICS

Activity 189
Solve the recurrence $a_n = 4a_{n-1} - 4a_{n-2}$. You did this already in Exercise 89 using the characteristic root technique, but there you had to be careful of the repeated root. How does that manifest itself here?

The point of the previous exercise is to notice how generating functions illustrate where the extra factor of n comes from in the general form $a_n = \alpha r^n + \beta n r^n$ you get for a closed formula when using the characteristic root technique on a recurrence that gives you a repeated root r. For any r, we have the power series

$$\frac{1}{(1-rx)^2} = 1 + 2(rx) + 3(rx)^2 + \cdots + (n+1)(rx)^n + \cdots,$$

using the extended binomial theorem and substituting rx for x. The coefficient of x^n is then $(n+1)r^n = r^n + nr^n$.

CATALAN NUMBERS.

Exercise 190
Recall the recurrence for the Catalan numbers is

$$C_n = \sum_{i=1}^{n-1} C_{i-1} C_{n-i}.$$

(a) Use $C_0 = 1$ and compute C_1, C_2, C_3, C_4, C_5.

(b) Show that if we use y to stand for the power series $\sum_{n=0}^{\infty} C_n x^n$, then we can find y by solving a quadratic equation. Find y. [Hint]

(c) Taylor's theorem from calculus tells us that the extended binomial theorem

$$(1+x)^r = \sum_{i=0}^{\infty} \binom{r}{i} x^i$$

holds for any number real number r, where $\binom{r}{i}$ is defined to be

$$\frac{P(r,i)}{i!} = \frac{r(r-1)\cdots(r-i+1)}{i!}.$$

Use this and your solution for y (note that of the two possible values for y that you get from the quadratic formula, only one gives an actual power series) to get a formula for the Catalan numbers. [Hint]

Summary of Generating Functions.

Sequence:	Power Series:	Generating Function:
$1, 1, 1, \ldots$	$1 + x + x^2 + x^3 + \cdots$	$\dfrac{1}{1-x}$
$1, 2, 4, 8, \ldots$	$1 + 2x + (2x)^2 + (2x)^3 + \cdots$	$\dfrac{1}{1-2x}$
$1, -1, 1, -1, \ldots$	$1 - x + x^2 - x^3 + \cdots$	$\dfrac{1}{1+x}$
$0, 3, 3^2, 3^3, \ldots$	$3x + (3x)^2 + (3x)^3 + \cdots$	$\dfrac{3x}{1-3x}$
$1, a, a^2, \ldots\ldots$	$1 + ax + (ax)^2 + \cdots$	$\dfrac{1}{1-ax}$
$1, 2, 3, 4, \ldots$	$1 + 2x + 3x^2 + 4x^3 + \cdots$	$\dfrac{1}{(1-x)^2}$
$0, 1, 2, 3, \ldots$	$x + 2x^2 + 3x^3 + \cdots$	$\dfrac{x}{(1-x)^2}$
$1, 0, 2, 0, 3, 0, \ldots$	$1 + 2x^2 + 3x^4 + \cdots$	$\dfrac{1}{(1-x^2)^2}$
$\binom{n}{0}, \binom{n}{1}, \binom{n}{2}, \ldots$	$\binom{n}{0} + \binom{n}{1}x + \binom{n}{2}x^2 + \cdots$	$(1+x)^n$
$\binom{2}{2}, \binom{3}{2}, \binom{4}{2}, \ldots$	$\binom{2}{2} + \binom{3}{2}x + \binom{4}{2}x^2 + \cdots$	$\dfrac{1}{(1-x)^3}$
$1, -2, 3, -4, \ldots$	$1 - 2x + 3x^2 - 4x^3 + \cdots$	$\dfrac{1}{(1+x)^2}$
$1, 2(2), 3(2^2), \ldots$	$1 + 2(2x) + 3(2x)^2 + \cdots$	$\dfrac{1}{(1-2x)^2}$
$1, 1, \tfrac{1}{2!}, \tfrac{1}{3!}, \ldots$	$1 + x + \tfrac{x^2}{2!} + \tfrac{x^3}{3!} + \cdots$	e^x

2.3 Partitions of Sets

Partitions are one of the core ideas in discrete mathematics. Recall that a **partition** of a set S is a collection of mutually disjoint subsets of S whose union is all of S. In other words, every element of S belongs to exactly one of the subsets of the partition. We call the subsets that make up the partition **blocks** or **parts** of the partition.

The integers can be naturally partitioned into evens and odds. In fact, this is just a specific case of partitioning the integers by their reminder when divided by a fixed constant; another example is that every integer has remainder 0, 1, or 2 when divided by 3.

Finite sets also have natural partitions: the set of all 4-bit strings can be partitioned according to weight:

$$B_0^4 \cup B_1^4 \cup B_2^4 \cup B_3^4 \cup B_4^4 = B^4.$$

Using the sum principle on these disjoint sets leads directly to the identity $\binom{4}{0} + \binom{4}{1} + \binom{4}{2} + \binom{4}{3} + \binom{4}{4} = 2^4$ (and of course the 4 can be replaced with any n here).

In fact, many of the identities we established in Section 1.2 using a combinatorial proof work precisely because we can count a set of outcomes all at once and also by counting the size of each block in a particular partition. For example, in Exercise 39 we proved that

$$\binom{n}{0}d_0 + \binom{n}{1}d_1 + \binom{n}{2}d_2 + \ldots + \binom{n}{n}d_n = n!$$

by partitioning the $n!$ permutations of $[n]$ into blocks according to how many are fixed in position while the rest are deranged.

Even the formula for binomial coefficients, $\binom{n}{k} = \frac{n!}{(n-k)!k!}$, comes from partitioning the k-permutations of $[n]$ into blocks according to which subset of $[n]$ is selected (and then $\binom{n}{k}$ is precisely the number of blocks).

In this section we will abstract this fundamental idea of partitions by one level. Instead of using particular partitions to answer counting questions, we will ask counting questions about partitions themselves. As we will see, this helps us solve yet more counting questions.

In Section 1.3 we considered some ways to distribute items to recipients. Most basic counting formulas can be thought of as counting the number of ways to distribute either distinct or identical items to distinct recipients. For example, distributing k distinct items to n distinct recipients can be done in n^k ways, if recipients can receive any number of items, or $P(n,k)$ ways if recipients can receive at most one item. If the items are identical, the corresponding number of ways to distribute them are $\left(\!\binom{n}{k}\!\right)$ and $\binom{n}{k}$.

What if the recipients are not distinct? Say we wish to distribute k books (either distinct or identical) to n identical boxes. This is a perfectly natural extension, but we will see the answer is not. First, let's get a feel for what this might look like for some small values of k and n.

Activity 191
Suppose you have 3 distinct books you want to put into 5 identical boxes.

(a) How many ways can you do this if each box can have at most one book?

(b) How many ways can you do this if any box can have any number of books? You might want to consider three cases: one, two, or three boxes are used. Assume we do not care about the order in which the books are placed inside boxes.

(c) Describe the outcomes we are counting in abstract mathematical terms. What sort of mathematical objects are we counting?

Activity 192
Suppose you have 3 identical books you want to put in 5 identical boxes.

(a) How many ways can you do this if each box can have at most one book?

(b) How many ways can you do this if any box can have any number of books? Again, you should consider three cases.

(c) What mathematical objects are we counting here?

Both of the previous two activities can be solved by counting partitions. The difference comes down to whether we partition a set of distinct objects or identical objects. It is not immediately clear how to model partitioning identical objects, and we will put this off until Section 2.4. Partitioning *distinct* objects simply means finding a partition of a set, the topic of the current section.

2.3.1 Stirling Numbers of the Second Kind

We would like to count the number of ways to partition a set.

Definition 2.3.1
Denote by $S(k, n)$ the number of partitions of $[k]$ into exactly n subsets. We call $S(k, n)$ a **Stirling number (of the second kind)**.

Note that we write $S(k, n)$ instead of $S(n, k)$ here because we try to use k for the number of elements being distributed and n for the number of recipients. When other books use $S(n, k)$, they mean the number of partitions of $[n]$ into exactly k blocks. This is the same definition as we give, but with our renamed variables the formulas we get might look different.

For example, consider how to partition $[3]$ into exactly two sets:

$$\{1, 2\}, \{3\} \qquad \{1, 3\}, \{2\} \qquad \{2, 3\}, \{1\}$$

and that is all, so $S(3, 2) = 3$. We do not care about the order the elements appear in each block, nor the order in which the blocks appear. Thus we see the Stirling numbers count the number of ways to distribute k distinct items to n identical recipients so that each recipient gets at least one item.

> **Activity 193**
> Get to know the Stirling numbers by finding some. List the set of partitions and count them.
>
> (a) Find $S(3, 1)$, $S(4, 1)$ and $S(k, 1)$.
>
> (b) Compute $S(2, 2)$, $S(3, 2)$ and $S(4, 2)$. Find a formula for $S(k, 2)$ and prove it is correct.
>
> (c) Compute $S(3, 3)$, and $S(4, 3)$.
>
> (d) Find formulas and give proofs for $S(k, k)$ and $S(k, k-1)$. [Hint]

We can arrange the Stirling numbers into a triangle (called **Stirling's second triangle**). The first five rows are shown below. The entries are indexed differently than in Pascal's triangle: the top 1 represents $S(1, 1)$, so for example, $S(5, 2) = 15$.

$$
\begin{array}{ccccccccc}
 & & & & 1 & & & & \\
 & & & 1 & & 1 & & & \\
 & & 1 & & 3 & & 1 & & \\
 & 1 & & 7 & & 6 & & 1 & \\
1 & & 15 & & 25 & & 10 & & 1
\end{array}
$$

How were these found? Sure, we could have written out all 25 of the partitions of $[5]$ into exactly 3 blocks, but we didn't. Could there be some way to get 25 from the entries above it? That is, what is the recurrence relation among the Stirling numbers?

> **Activity 194**
> Write down (if you haven't already) all 6 partitions of $[4]$ into 3 blocks. Break these into two cases by where 4 is: is 4 in solitary confinement (in a singleton set) or does he have a cellmate?

Now that you have some experience with listing the $S(4, 3)$ partitions of $[4]$, the next example will help us to generalize.

> **Example 2.3.2**
> Let's see how to take $S(4, 2) = 7$ and $S(4, 3) = 6$ to compute $S(5, 3) = 25$.
> First, here are the partitions of $[4]$ into 2 blocks (group A) or 3 blocks (group B), side by side:

Group A: Group B:

$\{1\}, \{2,3,4\}$ $\{1,2\}, \{3,4\}$ $\{3\}, \{1,2\}, \{4\}$
$\{2\}, \{1,3,4\}$ $\{1,3\}, \{2,4\}$ $\{2\}, \{1,3\}, \{4\}$
$\{3\}, \{1,2,4\}$ $\{2,3\}, \{1,4\}$ $\{2\}, \{1,4\}, \{3\}$
$\{4\}, \{1,2,3\}$ $\{1\}, \{2,3\}, \{4\}$
                                              $\{1\}, \{2,4\}, \{3\}$
                                              $\{1\}, \{3,4\}, \{2\}$

(Notice the $\binom{4}{2}$ ways of listing the middle column in group B.)

Now, to list the $S(5,3)$ partitions of $[5]$ into 3 parts, we need to add 5 somewhere. We can either

a. Let 5 be a singleton block, along with any of the partitions from group A; or

b. Let 5 join any of the blocks in any of the partitions of group B.

The second option can be accomplished in $3 \cdot 6$ ways. Thus $S(5,3) = 7 + 3 \cdot 6$.

Activity 195
Now generalize. In a partition of the set $[k]$, the number k is either in a block by itself, or it is not. Find a two variable recurrence for $S(k,n)$, valid for k and n larger than one. [Hint]

Exercise 196
Find a recurrence for the Lah numbers $L(k,n)$ similar to the one in Activity 195. [Hint]

Exercise 197
Extend Stirling's triangle enough to allow you to answer the following question and answer it. (Don't fill in the rows all the way; the work becomes quite tedious if you do. Only fill in what you need to answer this question.) A caterer is preparing three bag lunches for hikers. The caterer has nine different sandwiches. In how many ways can these nine sandwiches be distributed into three identical lunch bags so that each bag gets at least one?

We have often thought of counting problems as asking about how many functions there are from a k-elements set to an n-element set. The answer to this question is n^k for all functions and $P(n,k)$ for injective functions. What about surjective functions? There is a reason we haven't asked this question yet, but now we can at least get an expression for the number of surjections in terms of Stirling numbers.

Exercise 198
Given a function f from a k-element set K to an n-element set, we can define a partition of K by putting x and y in the same block of the partition if and only if $f(x) = f(y)$. How many blocks does the partition have if f is surjective? How is the number of functions from a k-element set *onto* an n-element set related to a Stirling number? Be as precise in your answer as you can. [Hint]

We do not have an explicit formula for either the number of surjections or for $S(k, n)$ yet, but note that if we could find either, we would now have both. We will see one approach to this soon.

Exercise 199
Each function from a k-element set K to an n-element set N is a function from K onto *some* subset of N. If J is a subset of N of size j, you know how to compute the number of functions that map onto J in terms of Stirling numbers. Suppose you add the number of functions mapping onto J over all possible subsets J of N. What simple value should this sum equal? Write the equation this gives you. [Hint]

Exercise 200
In how many ways can the sandwiches of Exercise 197 be placed into three distinct bags so that each bag gets at least one?

We will further investigate Stirling numbers in the rest of this section, but first, notice that we have not looked at the number of ways to partition a set into *any* number of blocks.

Definition 2.3.3
The total number of partitions of a k-element set is denoted by B_k and is called the k-th **Bell number**.

Example 2.3.4
There are five partitions of $[3]$:

$$\{1\}, \{2\}, \{3\}$$
$$\{1, 2\}, \{3\} \quad \{1, 3\}, \{2\} \quad \{1\}, \{2, 3\}$$
$$\{1, 2, 3\}$$

Note that $B_3 = S(3, 1) + S(3, 2) + S(3, 3) = 1 + 3 + 1$.

Activity 201

(a) Why is $B_k = \sum_{n=1}^{k} S(k,n)$, but $n^k \neq \sum_{n=1}^{k} S(k,n)n!$? Why is this a meaningful question? [Hint]

(b) Find a recurrence that expresses B_k in terms of B_n for $n < k$ and prove your formula correct in as many ways as you can. [Hint]

(c) Find B_k for $k = 1, 2, 4, 5, 6$.

The Bell numbers are interesting in their own right, and we will look at them more in Subsection 2.3.4.

2.3.2 Formulas for Stirling Numbers (of the second kind)

Here is a brief summary of what we have already discovered.

Directly from the definition, we see that $S(k,1) = 1$ and $S(k,k) = 1$. From these we can compute the remaining Stirling numbers using the recursion

$$S(k,n) = nS(k-1,n) + S(k-1,n-1).$$

This recursion is justified by dividing the partitions of $[k]$ into those which contain k as a singleton set (there are $S(k-1,n-1)$ of these) and those that don't (there are n choices for where k could be for each of the $S(k-1,n)$ partitions of $[k-1]$ into n blocks).

From this, we generated Stirling's second triangle.

```
              1
           1     1
        1     3     1
     1     7     6     1
  1    15    25    10    1
```

Exercise 202
Use the recursion to find the 6th row of the triangle.

Let's consider some ways to compute Stirling numbers directly.

While we might not have a nice closed formula for all Stirling numbers in terms of k and n, we can give closed formulas for those Stirling numbers close to the edges of the triangle. We have already considered some of these in Activity 193. To get back into the right mindset, start by reproving one of these.

Exercise 203
Prove $S(k, k-1) = \binom{k}{2}$. [Hint]

The proof of the formula above suggests that looking at the sizes of the blocks might be helpful. Let's see what this might look like with an example.

Activity 204
How many ways are there to partition $[5]$ into three sets?

(a) How many partitions of $[5]$ have two blocks of size 1 and one block of size 3?

(b) How many partitions of $[5]$ have one block of size 1 and two of size 2? [Hint]

(c) Are there any other types of partitions we need to consider? What is $S(5,3)$?

(d) Generalize! Find a formula for $S(k, k-2)$.

(e) A friend tells you that $S(k, k-2) = \binom{k}{3} + 3\binom{k}{4}$. Prove this is correct also. If this is the formula you found for the previous part, see the hint. [Hint]

Whenever we want to partition k items in n blocks, we can consider cases. We will represent each type of case with a **type vector** consisting of a sequence $(a_1, a_2, \ldots, a_k)$ where each a_i is the number of blocks that have size i. For example, in partitioning $[5]$ into 3 blocks, the previous activity first asked for the number of partitions with type vector $(2, 0, 1, 0, 0)$, two blocks of size 1 and one of size 3, and then for the number of partitions with type vector $(1, 2, 0, 0, 0)$, one block of size 1 and two of size 2. Note that $\sum_{i=1}^{k} a_i = n$ and $\sum_{i=1}^{k} i a_i = k$.

Activity 205

(a) How many partitions of $[6]$ are there into 3 blocks? List all the type vectors and the number of partitions for each.

(b) Generalize to find a formula for $S(k, k-3)$. Then clean up the formula so it is a linear combination of $\binom{k}{4}$, $\binom{k}{5}$, and $\binom{k}{6}$. [Hint]

Activity 206
In how many ways can we partition k items into n blocks so that we have a_i blocks of size i for each i? That is, given the type vector $(a_1, a_2, \ldots, a_k)$, how many partitions of $[k]$ into n blocks have that type vector? [Hint]

Activity 207
Describe how to compute $S(k, n)$ in terms of quantities given by the formula you found in Activity 206.

Activity 208
Explain why $S(k, 2) = 2^{k-1} - 1$. Then find another expression for $S(k, 2)$ using type vectors to establish an interesting identity.

Interlude: Multinomial Coefficients.

Recall that the Stirling numbers counted the number of ways to distribute 9 distinct sandwiches to three identical sandwich bags. This suggests a more realistic question.

Activity 209
In how many ways may the caterer distribute the nine sandwiches into three identical bags so that each bag gets exactly three? Answer this question. [Hint]

Activity 210
In how many ways can the sandwiches of Activity 209 be placed into *distinct* bags so that each bag gets exactly three?

This is an example of a **multinomial coefficient**. Although these are not directly related to Stirling numbers, we can use the ideas of type vectors to investigate them. Let's do that now.

Activity 211
In how many ways may we label the elements of a k-element set with n distinct labels (numbered 1 through n) so that label i is used j_i times? (If we think of the labels as $y_1, y_2, \ldots, y_n$, then we can rephrase this question as follows. How many functions are there from a k-element set K to a set $N = \{y_1, y_2, \ldots y_n\}$ so that y_i is the image of j_i elements of K?) This number is called a **multinomial coefficient** and denoted by

$$\binom{k}{j_1, j_2, \ldots, j_n}.$$

[Hint]

Activity 212
Explain how to compute the number of functions from a k-element set K to an n-element set N by using multinomial coefficients. [Hint]

Activity 213
Explain how to compute the number of functions from a k-element set K onto an n-element set N by using multinomial coefficients. [Hint]

Activity 214
What do multinomial coefficients have to do with expanding the kth power of a multinomial $x_1 + x_2 + \cdots + x_n$? This result is called the **multinomial theorem**. [Hint]

2.3.3 Identities with Stirling Numbers

The entries in the Pascal Triangle, $\binom{n}{k}$, satisfy the nice recursion $\binom{n}{k} = \binom{n-1}{k} + \binom{n-1}{k-1}$ and also have a closed formula: $\binom{n}{k} = \frac{n!}{k!(n-k)!}$. Activity 195 gives us a recursion for the Stirling numbers $S(k,n)$ and, unfortunately, the best we can do for a "closed" formula is the following.[6]

Theorem 2.3.5

$$S(k,n) = \frac{1}{n!}\left[n^k - \binom{n}{1}(n-1)^k + \binom{n}{2}(n-2)^k - \ldots + (-1)^{n-1}\binom{n}{n-1}1^k\right]$$

We will prove this following an approach due to Polya. The idea is to compare Stirling numbers to surjections, which we know how to count using the principle of inclusion and exclusion (see Subsection 2.1.4).

Activity 215
Suppose you wished to paint k houses and you have n different colors available. Each house will get a single color, but you want to ensure that each color is used at least once.

(a) Use the principle of inclusion and exclusion to write a formula for the number of ways to paint the houses. [Hint]

(b) What about the paint colors and the houses does $S(k,n)$ count? Why does this NOT answer the question? By what factor are we off?

(c) Conclude that a formula for the Stirling numbers is

$$S(k,n) = \frac{1}{n!}\sum_{s=0}^{n}(-1)^s\binom{n}{s}(n-s)^k.$$

Activity 216
What does Theorem 2.3.5 tell us about $S(k,2)$ (which we already know)? What do we get for $S(k,3)$?

Here is another identity involving Stirling numbers.

Theorem 2.3.6
$$x^k = S(k,1)x + S(k,2)x(x-1) + \cdots + S(k,k)x(x-1)\cdots(x-k+1)$$

Exercise 217
Prove Theorem 2.3.6. [Hint]

---

[6] This is not really a closed formula since the number of terms in the summation is not fixed.

It is in this form that James Stirling originally developed these numbers. $S(k,n)$ is used to convert from powers to binomial coefficients as shown in the following:

> **Theorem 2.3.7**
> $x^k = S(k,1)\binom{x}{1}1! + S(k,2)\binom{x}{2}2! + \ldots + S(k,k)\binom{x}{k}k!$

> **Exercise 218**
> Prove Theorem 2.3.7

As an example of the previous two theorems, consider how to write x^3. Using Theorem 2.3.6, we can write

$$x^3 = S(3,1)x + S(3,2)x(x-1) + S(3,3)x(x-1)(x-2) = x + 3x(x-1) + x(x-1)(x-2).$$

Recognizing that the factors that involve x look like a factorial, we can convert them into binomial coefficients, which is what Theorem 2.3.7 says:

$$x^3 = \binom{x}{1} + 3\binom{x}{2}2! + \binom{x}{3}3! = \binom{x}{1} + 6\binom{x}{2} + 6\binom{x}{3}.$$

Less we think this is nothing more than an interesting factoid, we can use it to establish the following.

> **Example 2.3.8**
> Prove that $1^3 + 2^3 + 3^3 + \cdots + n^3 = \left(\frac{n(n+1)}{2}\right)^2$. In other words, the square of any triangular number is the sum of cubes!
>
> **Solution.** We recognize that each cube on the left-hand side can be expressed as $\binom{x}{1} + 6\binom{x}{2} + 6\binom{x}{3}$. Thus
>
> $$1^3 + 2^3 + \cdots + n^3 = \sum_{x=1}^{n}\left(\binom{x}{1} + 6\binom{x}{2} + 6\binom{x}{3}\right)$$
> $$= \sum_{x=1}^{n}\binom{x}{1} + 6\sum_{x=1}^{n}\binom{x}{2} + 6\sum_{x=1}^{n}\binom{x}{3}.$$
>
> Now each of the three sums of binomial coefficients can be simplified using the Hockey Stick Theorem, to get
>
> $$= \binom{n+1}{2} + 6\binom{n+1}{3} + 6\binom{n+1}{4}$$
> $$= \binom{n+1}{2} + 6\left(\binom{n+1}{3} + \binom{n+1}{4}\right)$$
> $$= \binom{n+1}{2} + 6\binom{n+2}{4}$$
>
> where the last step uses the Pascal triangle recurrence.

Finally, write these last two binomial coefficients using the factorial formula and simplify:

$$\frac{n(n+1)}{2} + \frac{(n+2)(n+1)n(n-1)}{4} = \left(\frac{n(n+1)}{2}\right)^2.$$

The remainder of this section contains some additional activities related to the identities above and Stirling numbers in general.

Exercise 219
List the sequence of elements that are row sums of the Stirling Triangle.

Exercise 220
How many subsets $\{a,b,c\}$ are there of $\{2,3,4,\ldots\}$ such that $abc = 2 \cdot 3 \cdot 5 \cdot 7 \cdot 11 \cdot 13 \cdot 17$?

Exercise 221
Let $S = \{2,3,4,\ldots\}$. How many ordered triples (a,b,c) are there in $S \times S \times S$ such that $abc = 2 \cdot 3 \cdot 5 \cdot 7 \cdot 11 \cdot 13 \cdot 17$?

Exercise 222
Determine the following:

(a) a,b,c so that $n^4 = 24\binom{n}{4} + 6a\binom{n}{3} + 2b\binom{n}{2} + c\binom{n}{1}$.

(b) a,b,c,d so that $n^5 = 5!\binom{n}{5} + a\binom{n}{4} + b\binom{n}{3} + c\binom{n}{2} + d\binom{n}{1}$.

Exercise 223
Express $1^4 + 2^4 + 3^4 + \cdots + n^4$ as a polynomial in n.

Exercise 224
Why is $4^k - 4 \cdot 3^k + 6 \cdot 2^k - 4$ always divisible by 24?

2.3.4 Bell Numbers

The Bell number, B_k, denotes the number of ways that a set of k objects can be partitioned into nonempty subsets. We have already seen how to partition a set of k objects into exactly n subsets: $S(k,n)$. So then $B_k = \sum_{n=1}^{k} S(k,n)$. We take $B_0 = 1$.

The sequence of Bell numbers starts

$$1, 1, 2, 5, 15, 52, 203, 877, 4140, 21147, 115975, \ldots$$

Notice that the first four numbers look like Catalan numbers. In fact, one of the many things the Catalan numbers count are *non-crossing* set partitions.[7] So if C_n counts only some of the set partitions, then it must be that $C_n \leq B_n$ for all n.

The Bell numbers can be generated by constructing what is called the Bell Triangle. To construct this triangle, begin with a 1 at the top and a 1 below it. Add these two numbers together and put the sum 2, to the right of the 1 in the second column. This 2 is also the first entry of the third row. The second entry in the third row is found by adding the 2 to the number 1 above it. This sum is 3 and goes to the right of the 2. The 3 is now below a 2. Adding these two numbers produces the last number, 5, in the third row. Since 5 has no number above it, the third row is complete. 5 now becomes the first number in the fourth row and the process continues.

```
1
1    2
2    3    5
5    7    10   15
15   20   27   37   52
52   67   87   114  151  203
203  255  322  409  523  674  877
```

To summarize, construction of the triangle follows two basic rules:

1. The last number of each row is the first number of the next row.

2. All other numbers are found by adding the number to the left of the missing number to the number directly above this same number.

When the triangle is extended, as above, the Bell Numbers are found down the first column as well as along the outside diagonal.

Exercise 225
Why does this triangle construction give you the Bell numbers? Explain in terms of set partitions.

As with Pascal's Triangle, the Bell triangle has several interesting properties.

Exercise 226
If the sum of a row is added to the Bell number at the end of that row, the next Bell number is obtained. For example, the sum of the fourth row plus the Bell number at the end of the row: $15 + 20 + 27 + 37 + 52 + 52$ is 203, the next Bell number. Why does this happen?

Exercise 227
Another pattern you might notice: the numbers of the second diagonal $1, 3, 10, 37, 151, 674, \ldots$ are the sums of the horizontal rows. Prove this!

---

[7]A set partition is said to be **non-crossing** as long as whenever a, b belong to one block and c, d belong to another, we have that a and b are both either larger or smaller than both of c and d.

Rotating the triangle slightly creates a *difference* triangle analogous to Pascal's Triangle. The entries that are formed recursively by adding in Pascal's Triangle are now differences of the two numbers above them in the Bell triangle.

```
1     2     5      15      52       203      877  ...
   1     3     10      37      151      674  ...
      2     7      27      114      523  ...
         5      20      87      409  ...
            15      67      322  ...
               52      255  ...
                  203  ...
```

Exercise 228
Explain why the differences work as they do in the rotated triangle.

Finally, rewriting the Bell triangle recursively indicates a nice connection of the Bell Numbers to Pascal's triangle.

$B_0 = B_1$
$B_1 \quad B_0 + B_1 = B_2$
$B_2 \quad B_1 + B_2 \quad B_0 + 2B_1 + B_2 = B_3$
$B_3 \quad B_2 + B_3 \quad B_1 + B_2 + B_2 + B_3 \quad B_0 + 3B_1 + 3B_2 + B_3 = B_4$

This pattern suggests the $(n+1)^{\text{th}}$ Bell Number can be represented recursively by $B_{n+1} = \binom{n}{0}B_0 + \binom{n}{1}B_1 + \binom{n}{2}B_2 + \ldots + \binom{n}{n}B_n$. The coefficients of this equation are the entries of the nth row of Pascal's triangle.

Exercise 229
Prove the recurrence

$$B_{k+1} = \binom{k}{0}B_0 + \binom{k}{1}B_1 + \binom{k}{2}B_2 + \ldots + \binom{k}{k}B_k$$

The Bell numbers are related to the Stirling numbers and can be defined as the sum of the Stirling numbers of the second kind. That is, $B_k = S(k, 1) + S(k, 2) + \cdots + S(k, k)$, where $S(k, n)$ represents the number of ways of grouping k elements into n subsets. So $B_3 = S(3, 1) + S(3, 2) + S(3, 3) = 1 + 3 + 1 = 5$. Since the Bell numbers count the partitions of a set of elements, they are used in prime-number theory to enumerate the number of ways to factor a number with distinct prime factors. For example, 42 has three distinct prime factors: 2, 3, and 7. Since $B_3 = 5$, we know there are five ways of factoring 42. These are $2 \times 3 \times 7$, 2×21, 3×14, 6×7, and 42. So the number 210, which has 4 distinct factors, 2, 3, 5, and 7, can be factored in $B_4 = 15$ ways.

The Bell numbers can be used to model many real-life situations. For example, the number of different ways two people can sleep in unlabeled twin beds is $B_2 = 2$ ways: They can sleep in the same bed or in separate beds. The number of ways of serving a dinner consisting of three items, such as a salad, bread, and fish is $B_3 = 5$ ways: each could be served on a separate plate, salad and bread could be on one plate and fish on another, salad and fish on one plate and bread on another, or all three items could be on the same plate. This example serves as a model for the five ways three people can occupy three unlabeled beds, the five ways three prisoners can be handcuffed together, the five ways three nations can be form alliances, or any situation of partitioning three distinct elements into non-empty subsets.

One interesting application of Bell numbers is in counting the number of rhyme schemes possible for a stanza in poetry. There are $B_2 = 2$ possibilities for a two-line stanza: the lines can either rhyme or not rhyme. The possible rhyme schemes of a three-line stanza can be described as aaa, aab, aba, abb, and abc: thus there are 5 or B_3 possible rhyme schemes. The Japanese used diagrams depicting possible rhyme schemes of a five-line stanza $B_5 = 52$ as early as 1000 A.D. in the *Tale of the Genji* by Lady Shikibu Murasaki.

Exercise 230
List all 15 rhyme schemes for a four-line stanza. Identify the $C_4 = 14$ such schemes that are "planar" (and say what this means).

Exercise 231
List all 15 set partitions of $[4]$. Identify the $C_4 = 14$ such partitions that are "non-crossing" (and say what this means).

We conclude with an interesting exercise.

Exercise 232
The set $\{u_0, u_1, u_2, \ldots\}$ where $u_0 = 1$, $u_1 = x$, $u_2 = x(x-1)$, $u_3 = x(x-1)(x-2)$, etc., is a basis for the vector space V of all polynomials with real coefficients. Let $P(x) = c_0 u_0 + c_1 u_1 + c_2 u_2 + \cdots$ be any element of V, and define the functional L as follows: $L[P(x)] = c_0 + c_1 + \cdots$.

1. Prove that $L[P(x) + Q(x)] = L[P(x)] + L[Q(x)]$.

2. Prove that $L[rP(x)] = rL[P(x)]$ for $r \in \mathbb{R}$ (the Reals).

3. Prove that $B_n = L[x^n]$.

4. Prove that $L[x^{n+1}] = L[(x+1)^n]$.

5. Derive the recursion in Exercise 229 using the previous part.

2.4 Partitions of Integers

In Example 1.3.4, we counted the *compositions* of an integer n, by counting the number of solutions to the equation $x_1 + x_2 + \cdots + x_k = n$ where each x_i is a positive integer. Put another way, we asked how many *lists* of k positive integers have sum n. The order in which we listed the sum mattered. What if it didn't?

2.4.1 Partition numbers

A multiset of positive integers that add to n is called a **partition** of n. Thus the partitions of 3 are 1+1+1, 1+2 (which is the same as 2+1) and 3. The number of partitions of k is denoted by the **partition number** $p(k)$; in computing the partitions of 3 we showed that $p(3) = 3$. It is traditional to use Greek letters like λ to stand for partitions; we might write $\lambda = 1, 1, 1$, $\gamma = 2, 1$ and $\tau = 3$ to stand for the three partitions we just described. We also write $\lambda = 1^3$ as a shorthand for $\lambda = 1, 1, 1$, and we write $\lambda \dashv 3$ as a shorthand for "λ is a partition of three."

> **Example 2.4.1**
> There are 5 partitions of 4 (so $p(4) = 5$). They are $4 = 1 + 1 + 1 + 1$, $4 = 2 + 1 + 1$, $4 = 2 + 2$, $4 = 3 + 1$, and $4 = 4$.

> **Activity 233**
> Write out all partitions of 5, to compute $p(5)$.

A **partition of the integer k into n parts** is a multiset of n positive integers that add to k. We use $p_n(k)$ to denote the number of partitions of k into n parts. Thus $p_n(k)$ is the number of ways to distribute k identical objects to n identical recipients so that each gets at least one.

> **Example 2.4.2**
> Here are the partitions of 7 into two parts: $6 + 1$, $5 + 2$, $4 + 3$. Thus $p_2(7) = 3$.

> **Activity 234**
> Find $p_3(6)$ by finding all partitions of 6 into 3 parts. What does this say about the number of ways to put six identical apples into three identical bags so that each bag has at least one apple?

> **Exercise 235**
> How many solutions are there in the positive integers to the equation $x_1 + x_2 + x_3 = 7$ with $x_1 \geq x_2 \geq x_3$?

Exercise 236
Explain the relationship between partitions of k into n parts and lists $x_1, x_2, \ldots, x_n$ of positive integers with $x_1 + x_2 + \cdots + x_n = k$ and $x_1 \geq x_2 \geq \ldots \geq x_n$. Such a representation of a partition is called a **decreasing list** representation of the partition.

Exercise 237
Show that $p_n(k)$ is at least $\frac{1}{n!}\binom{k-1}{n-1}$.   [Hint]

Exercise 238
Describe the relationship between partitions of k and lists of vectors $(x_1, x_2, \ldots, x_n)$ such that $x_1 + 2x_2 + \ldots kx_k = k$. Such a representation of a partition is called a **type vector** representation of a partition, and it is typical to leave the trailing zeros out of such a representation; for example $(2, 1)$ stands for the same partition as $(2, 1, 0, 0)$. What is the decreasing list representation for this partition, and what number does it partition?

Exercise 239
How does the number of partitions of k relate to the number of partitions of $k + 1$ whose smallest part is one?   [Hint]

When we write a partition as $\lambda = \lambda_1, \lambda_2, \ldots, \lambda_n$, it is customary to write the list of λ_is as a decreasing list. When we have a type vector $(t_1, t_2, \ldots, t_m)$ for a partition, we write either $\lambda = 1^{t_1} 2^{t_2} \cdots m^{t_m}$ or $\lambda = m^{t_m}(m-1)^{t_{m-1}} \cdots 2^{t_2} 1^{t_1}$. Henceforth we will use the second of these. When we write $\lambda = \lambda_1^{i_1} \lambda_2^{i_2} \cdots \lambda_n^{i_n}$, we will assume that $\lambda_i > \lambda_i + 1$.

It is remarkable that there is no known formula for $p_n(k)$, nor is there one for $p(k)$. We have computed a few values of $p_n(k)$ by listing partitions. If we continue to do so, we could build a triangle of partition numbers as follows.

$k\backslash n$	1	2	3	4	5	6	7
1	1						
2	1	1					
3	1	1	1				
4	1	2	1	1			
5	1	2	2	1	1		
6	1	3	3	2	1	1	
7	1	3	4	3	2	1	1

Figure 2.4.3 The triangle of partition numbers through 7.

So for example, $p_3(7) = 4$: $(5, 1, 1)$, $(4, 2, 1)$, $(3, 3, 1)$, and $(3, 2, 2)$. It sure would be nice to find the next row without having to list out all the partitions. Perhaps we could use some sort of recursion!

With the binomial coefficients, with Stirling numbers of the second kind, and with the Lah numbers, we were able to find a recurrence by asking what happens to our subset, partition, or broken permutation of a set S of numbers if we remove the largest element of S. Thus it is natural to look for a recurrence to count the number of partitions of k into n parts by doing something similar. Unfortunately, since we are counting distributions in which all the objects are identical, there is no way for us to identify a largest element. However if we think geometrically, we will be able to "see" how to get smaller partitions from larger.

2.4.2 Using Geometry

The decreasing list representation of partitions leads us to a handy way to visualize partitions. Given a decreasing list $(\lambda_1, \lambda_2, \ldots \lambda_n)$, we draw a figure made up of rows of dots that has λ_1 equally spaced dots in the first row, λ_2 equally spaced dots in the second row, starting out right below the beginning of the first row and so on. Equivalently, instead of dots, we may use identical squares, drawn so that a square touches each one to its immediate right or immediately below it along an edge. See Figure 2.4.4 for examples. The figure we draw with dots is called the **Ferrers diagram** of the partition; sometimes the figure with squares is also called a Ferrers diagram; sometimes it is called a **Young diagram**. At this stage it is irrelevant which name we choose and which kind of figure we draw; in more advanced work the squares are handy because we can put things like numbers or variables into them. From now on we will use squares and call the diagrams Young diagrams.

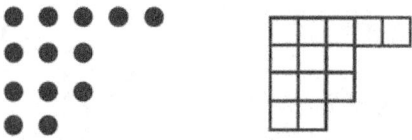

Figure 2.4.4 The Ferrers and Young diagrams of the partition (5,3,3,2)

Activity 240

We will soon start manipulating Young diagrams, so having some examples will be useful. Keep these in a safe place.

(a) Draw all the Young diagrams for the partitions of 4 and 5, grouping (i.e., partitioning) them by the number of parts.

(b) Draw all $p_2(6)$ Young diagrams for the partitions of 6 into 2 parts.

(c) Draw all $p_3(7)$ Young diagrams for the partitions of 7 into 3 parts.

(d) Draw all $p_4(9)$ Young diagrams for the partitions of 9 into 4 parts.

The natural way to get a partition of a smaller integer from a partition of n is to remove some of the squares from the Young diagram. We lots of choices here.

1. We could remove the entire top row.

2. We could remove the entire left-most column.

3. We could remove a single box somewhere. But from where?

Removing the top row corresponds to removing the largest part of the partition. Removing the left-most column corresponds to subtracting 1 from each part. Removing a single box will reduce one part by 1. Some of these might be more useful than others.

Activity 241
Let's see what happens when we remove the top row and whether this is helpful.

(a) Look at your partitions of 7 into 3 parts. Which partitions of what into how many parts do you get if you remove the top row? Looking at the partition number triangle, does this allow you to calculate $p_3(7) = 4$?

(b) Not so fast! Repeat the previous task with your partitions of 9 into 4 parts. What goes wrong here?

Activity 242
Now consider what happens when we remove the left-most column.

(a) Remove the left most column from each of your partitions of 7 into 3 parts. Which partitions of what into how many parts do you get? Comparing to the partition number triangle, does this allow you to compute $p_3(7) = 4$? [Hint]

(b) Repeat the previous task with partitions of 9 into 4 parts. Do we run into the same problem as when we removed the top row?

The previous activity suggests a recurrence for $p_n(k)$. We might guess

$$p_n(k) = \sum_{i=1}^{n} p_i(k-n) = p_1(k-n) + p_2(k-n) + \cdots + p_n(k-n).$$

For example,

$$p_4(9) = p_1(5) + p_2(5) + p_3(5) + p_4(5) = 1 + 2 + 2 + 1 = 6$$

(note we do not add the entire row, just up until the column we are interested in).

Exercise 243
Complete the 8th row of the partition number triangle using the recurrence.

Exercise 244
Prove the recurrence for $p_n(k)$ is correct. [Hint]

Wikipedia (and almost every other textbook) gives another recursion for $p_n(k)$:
$$p_n(k) = p_n(k-n) + p_{n-1}(k-1).$$

Let's call this the **short recursion for partition numbers** (and we will call our original recurssion the "long" one).

Activity 245
Does the short recursion really work?

(a) Verify the short recursion using numbers in the triangle. In particular, check $p_4(9)$.

(b) Use your Young diagrams you drew above to see what is going on with the short recursion. What geometric operations do the two parts of the sum represent?

(c) Generalize your observations in the previous task to explain why the short recursion is correct.

Activity 246
Suppose we wanted to compute $p_5(13)$. Since we have the 8th row of the partition number triangle, we should be able to do this with the long recursion. What if we wanted to use the short recursion?

Use the short recursion to compute $p_5(13)$. If you get a partition number that you don't yet know, use the short recursion to compute that as well (you know, the way you use recursions).

As you see above, the short recursion is nothing more than grouping all but the first term in the sum into a single partition number, using the long recursion again. This should make sense based on the geometry of the Young digrams.

MORE GEOMETRY.

Our goal is to use geometric reasoning about these diagrams (we will generally use the Young diagram with squares) to uncover properties of partitions.

Activity 247
Draw the Young diagram of the partition (4,4,3,1,1). Describe the geometric relationship between the Young diagram of (5,3,3,2) and the Young diagram of (4,4,3,1,1). [Hint]

Activity 248
The partition $(\lambda_1, \lambda_2, \ldots, \lambda_n)$ is called the **conjugate** of the partition $(\gamma_1, \gamma_2, \ldots, \gamma_m)$ if we obtain the Young diagram of one from the Young diagram of the other by flipping one around the line with slope -1 that extends the diagonal of the top left square. (Equivalently, simply inter-

changing rows and columns, like taking the transpose of a matrix.) See Figure 2.4.5 for an example.

Figure 2.4.5 The Young diagram for the partition (5,3,3,2) and its conjugate.

What is the conjugate of (4,4,3,1,1)? How is the largest part of a partition related to the number of parts of its conjugate? What does this tell you about the number of partitions of a positive integer k with largest part m? [Hint]

Activity 249
A partition is called **self-conjugate** if it is equal to its conjugate. Find a relationship between the number of self-conjugate partitions of k and the number of partitions of k into distinct odd parts. [Hint]

Example 2.4.6
Restricting the number or type of parts often leads to interesting relationships.

We must be careful about what we mean. For example, here are three different sorts of "even" partitions for $k = 10$:

Even parts:	Even number of parts:	Parts with even multiplicity:
(10)	(9, 1)	(5, 5)
(8, 2)	(8, 2)	(4, 4, 1, 1)
(6, 4)	(7, 3)	(3, 3, 2, 2)
(6, 2, 2)	(6, 4)	(3, 3, 1, 1, 1, 1)
(4, 4, 2)	(5, 5)	(2, 2, 2, 2, 1, 1)
(4, 2, 2, 2)	(4, 3, 2, 1)	(2, 2, 1, 1, 1, 1, 1, 1)
(2, 2, 2, 2, 2)		(1, 1, 1, 1, 1, 1, 1, 1, 1, 1)

Activity 250
Explain the relationship between the number of partitions of k into even parts and the number of partitions of k into parts of even multiplicity, i.e. parts which are each used an even number of times as in (3,3,3,3,2,2,1,1). [Hint]

Exercise 251
Show that the number of partitions of k into four parts equals the number of partitions of $3k$ into four parts of size at most $k-1$ (or $3k-4$ into four parts of size at most $k-2$ or $3k-4$ into four parts of size at most k). [Hint]

Exercise 252
The idea of conjugation of a partition could be defined without the geometric interpretation of a Young diagram, but it would seem far less natural without the geometric interpretation. Another idea that seems much more natural in a geometric context is this. Suppose we have a partition of k into n parts with largest part m. Then the Young diagram of the partition can fit into a rectangle that is m or more units wide (horizontally) and n or more units deep. Suppose we place the Young diagram of our partition in the top left-hand corner of an m' unit wide and n' unit deep rectangle with $m' \geq m$ and $n' \geq n$, as in Figure 2.4.7.

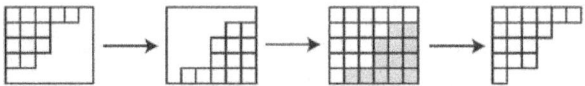

Figure 2.4.7 To complement the partition $(5, 3, 3, 2)$ in a 6 by 5 rectangle: enclose it in the rectangle, rotate, and cut out the original Young diagram.

(a) Why can we interpret the part of the rectangle not occupied by our Young diagram, rotated in the plane, as the Young diagram of another partition? This is called the **complement** of our partition in the rectangle.

(b) What integer is being partitioned by the complement?

(c) What conditions on m' and n' guarantee that the complement has the same number of parts as the original one? [Hint]

(d) What conditions on m' and n' guarantee that the complement has the same largest part as the original one? [Hint]

(e) Is it possible for the complement to have both the same number of parts and the same largest part as the original one?

(f) If we complement a partition in an m' by n' box and then complement that partition in an m' by n' box again, do we get the same partition that we started with?

Exercise 253
Suppose we take a partition of k into n parts with largest part m, complement it in the smallest rectangle it will fit into, complement the result in the smallest rectangle it will fit into, and continue the process until we get the partition 1 of one into one part. What can you say about the partition with which we started? [Hint]

2.4.3 Partitions into distinct parts

Often $q_n(k)$ is used to denote the number of partitions of k into distinct parts, that is, parts that are different from each other.

> **Example 2.4.8**
> Show that the number of partitions of 7 into three parts equals the number of partitions of 10 into three distinct parts.
>
> **Solution.** To get a feel for this, list them out:
>
> $$(5,1,1) \qquad (7,2,1)$$
> $$(4,2,1) \qquad (6,3,1)$$
> $$(3,3,1) \qquad (5,4,1)$$
> $$(3,2,2) \qquad (5,3,2)$$
>
> There are four of each sort, which shows what we want. But what is the connection? It's to add 2 1 0!
>
> Given a partition λ of 7 in decreasing list form $\lambda_1, \lambda_2, \lambda_3$, if we add 0 to λ_3, 1 to λ_2 and 2 to λ_1 the resulting partition of 10 has distinct parts. If we take a partition λ' of 10 with distinct parts, then $\lambda'_1 \geq \lambda'_2 + 1$, $\lambda'_1 \geq \lambda'_3 + 2$, and $\lambda'_2 \geq \lambda'_3 + 1$. Therefore if we subtract 2 from λ'_1 to get λ_1, subtract 1 from λ'_2 to get λ_2 and let $\lambda_3 = \lambda'_3$, then $\lambda_1, \lambda_2, \lambda_3$ is the decreasing list representation of a partition of $10 - 3 = 7$. Thus there is a bijection between partitions of 7 into three parts and partitions of 10 into three distinct parts.

> **Exercise 254**
> Show that the number of partitions of 10 into three parts equals the number of partitions of 13 into three distinct parts.

> **Exercise 255**
> There is a relationship between $p_n(k)$ and $q_n(m)$ for some other number m. Find the number m that gives you the nicest possible relationship. [Hint]

> **Activity 256**
> Find a recurrence that expresses $q_n(k)$ as a sum of $q_m(k-n)$ for appropriate values of m. [Hint]

The following exercise is interesting, but difficult. The adventurous reader can try it directly here. We will establish the same result using generating functions in the next section.

> **Exercise 257**
> Show that the number of partitions of k into distinct parts equals the number of partitions of k into odd parts. [Hint]

Here is another challenging exercise.

Exercise 258
Euler showed that if $k \neq \frac{3j^2+j}{2}$, then the number of partitions of k into an even number of distinct parts is the same as the number of partitions of k into an odd number of distinct parts. Prove this, and in the exceptional case find out how the two numbers relate to each other. [Hint]

And finally, a much simpler, purely combinatorial result.

Exercise 259
Show that
$$q_n(k) \leq \frac{1}{n!}\binom{k-1}{n-1}.$$
[Hint]

2.4.4 Using Generating Functions

I this final section we will see that generating functions are especially well suited for the study of integer partitions. We will start with a classic example.

Activity 260
If we have five identical pennies, five identical nickels, five identical dimes, and five identical quarters, give the picture enumerator for the combinations of coins we can form and convert it to a generating function for the number of ways to make k cents with the coins we have. Do the same thing assuming we have an unlimited supply of pennies, nickels, dimes, and quarters. [Hint]

Activity 261
In Activity 260 we found the generating function for the number of partitions of an integer into parts of size 1, 5, 10, and 25.

(a) Give the generating function for the number partitions of an integer into parts of size one through ten. [Hint]

(b) Give the generating function for the number of partitions of an integer k into parts of size at most m, where m is fixed but k may vary. Notice this is the generating function for partitions whose Young diagram fits into the space between the line $x = 0$ and the line $x = m$ in a coordinate plane. (We assume the boxes in the Young diagram are one unit by one unit.) [Hint]

Activity 262

In Task 261.b you gave the generating function for the number of partitions of an integer into parts of size at most m. Can you see why this is also the generating function for partitions of an integer into at most m parts. [Hint]

Example 2.4.9
The generating function for the number of partitions of an integer into parts of any size is

$$\frac{1}{(1-x)(1-x^2)(1-x^3)\cdots} = (1+x+x^2+x^3+\cdots)(1+x^2+x^4+\cdots)(1+x^3+\cdots)\cdots.$$

The coefficient of x^4 is 5. Where does this come from exactly?

First notice that the x^4 in the second factor is really $x^{2 \cdot 2}$. That is, while the first factor's x^4 represents using four 1's, the second factor's x^4 comes from $(x^2)^2$ so represents using two 2's. The fourth factor's x^4 means one 4.

Here are all the ways of forming x^4 using the first four factors (we would never need anything beyond this), and their corresponding partition.

$$1 \cdot 1 \cdot 1 \cdot x^4 \leftrightarrow (4)$$
$$x \cdot 1 \cdot x^3 \cdot 1 \leftrightarrow (3,1)$$
$$1 \cdot x^4 \cdot 1 \cdot 1 \leftrightarrow (2,2)$$
$$x^2 \cdot x^2 \cdot 1 \cdot 1 \leftrightarrow (2,1,1)$$
$$x^4 \cdot 1 \cdot 1 \cdot 1 \leftrightarrow (1,1,1,1)$$

Exercise 263
Illustrate the partitions of 5 and their correspondence to the ways to get x^5 in the generating function for the number of partitions of an integer. [Hint]

Example 2.4.10
The generating function for the number of partitions of an integer in which each part is even is

$$\frac{1}{(1-x^2)(1-x^4)(1-x^6)\cdots}$$

Example 2.4.11
Explain why the generating function for the number of partitions of an integer into exactly m parts is

$$\frac{x^m}{(1-x)(1-x^2)\cdots(1-x^m)}.$$

Solution. By considering a Young diagram and its compliment, we saw that there is a one to one correspondence between partitions that have exactly m parts and partitions that have m as their largest part.

The generating function for the number of partitions of integers having *largest* part m is
$$\frac{1}{(1-x)(1-x^2)\cdots(1-x^m)}.$$
Thus this is also the number of partitions of integers having *at most* m parts. Now subtract those having at most $m-1$ parts:
$$\frac{1}{(1-x)\cdots(1-x^m)} - \frac{1}{(1-x)\cdots(1-x^{m-1})} = \frac{x^m}{(1-x)\cdots(1-x^m)}.$$

Exercise 264

Show that the generating function for the number of partitions of integers into *exactly m distinct* parts is
$$\frac{x^{\frac{m(m+1)}{2}}}{(1-x)(1-x^2)\cdots(1-x^m)}.$$

Example 2.4.12

Let's illustrate these last few results with $m = 3$.

First,
$$\frac{1}{(1-x)(1-x^2)(1-x^3)} = 1 + x + 2x^2 + 3x^3 + 4x^5 + 7x^6 + \cdots.$$

This shows that the number of partitions of $n = 6$ into at most 3 parts is 7. The are

$$(6)$$
$$(5,1)$$
$$(4,2)$$
$$(4,1,1)$$
$$(3,3)$$
$$(3,2,1)$$
$$(2,2,2).$$

Now multiply the series by x^3. The term $3x^6$ shows that there are 3 partitions of $n = 6$ into exactly 3 parts.

Finally, multiply instead by $x^{\frac{m(m+1)}{2}} = x^6$ (when $m = 3$) to get $x^6 + x^7 + 2x^8 + 3x^9 + 4x^{10} + \cdots$. This shows that only one of the partitions of $n = 6$ into 3 parts has its parts distinct. Further, there are 4 partitions of $n = 10$ into exactly 3 distinct parts:

$$(7,2,1)$$

$$(6, 3, 1)$$
$$(5, 4, 1)$$
$$(5, 3, 2).$$

Exercise 265
Give the generating function for the number of partitions of n with each of the following restrictions

(a) Each part is used at most once (i.e., a partition into distinct parts).

(b) No part is larger than 3.

(c) All parts are odd.

(d) All parts are distinct and odd.

(e) Only even parts may be repeated.

(f) No part appears more than twice.

(g) All parts are in the set $\{3, 8\}$. (This might remind you of a stamp problem.)

Exercise 266
Use generating functions to explain why the number of partitions of an integer in which each part is used an even number of times equals the number of partitions of an integer in which each part is even. [Hint]

Exercise 267
Use the fact that
$$\frac{1-x^{2i}}{1-x^i} = 1 + x^i$$
and the generating function for the number of partitions of an integer into distinct parts to show how the number of partitions of an integer k into distinct parts is related to the number of partitions of an integer k into odd parts. [Hint]

Exercise 268
Verify that the number of partitions of $n = 5$ is equal to the number of partitions of 10 into exactly 5 parts. Generalize.

Exercise 269
Write down the generating function for the number of ways to partition an integer into parts of size no more than m, each used an odd number of times. Write down the generating function for the number of partitions of an integer into parts of size no more than m, each used an even number of times. Use these two generating functions to get a relationship between the two sequences for which you wrote down the generating functions.

[Hint]

Exercise 270
Show algebraically that

$$(1+x)(1+x^2)(1+x^4)(1+x^8)\cdots = \frac{1}{(1-x)(1-x^3)(1-x^5)\cdots}.$$

What does this identity tell you about partitions of integers?

Exercise 271

(a) Prove that
$$\frac{1}{1-x} = (1+x+x^2+\cdots+x^9)(1+x^{10}+\cdots+x^{90})(1+x^{100}+\cdots+x^{900})\cdots.$$

What does this tell us about numbers?

(b) Prove that every non negative integer n has a unique binary representation.

(c) Find a simple expression for
$$(1+x+x^2)(1+x^3+x^6)(1+x^9+x^{18})(1+x^{27}+x^{54})\cdots.$$

How can you interpret this result?

Chapter 3
Graph Theory

To start our journey into discrete mathematics, let's consider a few puzzles.

Puzzle 272
In the time of Euler, in the town of Königsberg in Prussia, there was a river containing two islands. The islands were connected to the banks of the river by seven bridges (as seen below). The bridges were very beautiful, and on their days off, townspeople would spend time walking over the bridges. As time passed, a question arose: was it possible to plan a walk so that you cross each bridge once and only once? Euler was able to answer this question. Are you?

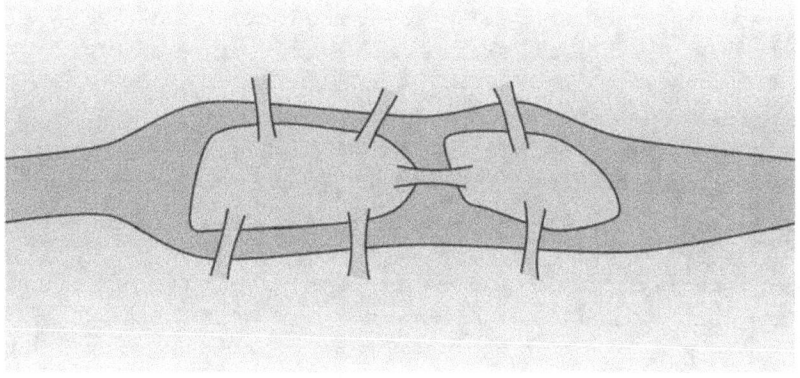

Puzzle 273
Professor Nefario has invited 25 unsuspecting discrete math students to his estate for an evening of escape room fun. Some of these students are already friends, but no student is friends with more than 3 other students. The Professor wants to ensure that no friends are sent to the same escape room. Can he be sure that four escape rooms will be enough?

Puzzle 274
Is it possible that among the 25 students, everyone has *exactly* three friends?

Puzzle 275
Take a regular deck of 52 cards, shuffle well, and deal the cards into 13 piles. Will it always be possible to select one card from each pile so that the 13 cards you select all have different values (ace, 2, 3, etc.)?

All the puzzles above have a feature in common: they involve the ways in which individual (discrete) objects can be related to each other. This is what graph theory studies, and we will see that each of the puzzles can be solved with the tools of the subject. Before we start developing those tools, take a moment to see why graphs are applicable to these problems.

In the bridges of Königsberg puzzle, all that really matters is which land masses are connected by bridges, and how many bridges connect them. The length of the bridges and size of the land masses is unimportant. So we can represent this puzzle with a graph like this:

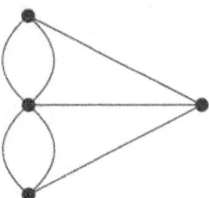

Now we must decide whether it is possible to travel around that graph in a way that each edge is crossed exactly once.

When it comes to friendship, if we assume that it is always reciprocated, we can represent the people as vertices and connect two vertices with an edge if the people they represent are friends. We can ask whether it is possible to have such a graph in which each vertex is incident to exactly three edges or whether it is possible to "color" the vertices with just four colors so that no pair of same colored vertices are adjacent (connected by an edge).

If we have thirteen piles of cards, we represent each pile as a vertex and each value as a vertex, and connect a pile vertex to a value vertex precisely when that value is in the pile. We could then ask whether we could select a collection of thirteen edges that do not share any single vertex.

These puzzles give you just a small glimpse into the myriad of applications graph theory can address. For us, we will use our study of graph theory to see how reasoning in discrete mathematics works and as an excuse to remind ourselves of the basic mathematical objects and proof techniques required to study the subject in general.

3.1 Definitions

Activity 276
For now, let's say that a **graph** is simply a collection of vertices, some of which are connected by an edge. Draw all graphs that have exactly five vertices and six edges. How many are there? How do you know?

Activity 277

Which (if any) of the graphs below are the same?

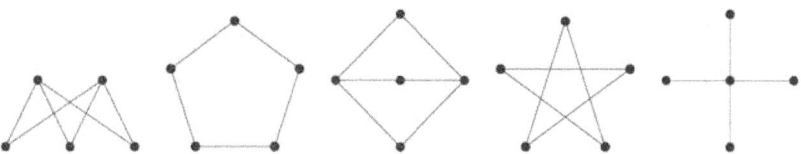

The graphs above are unlabeled. Usually we think of a graph as having a specific set of vertices. Which (if any) of the graphs below are the same?

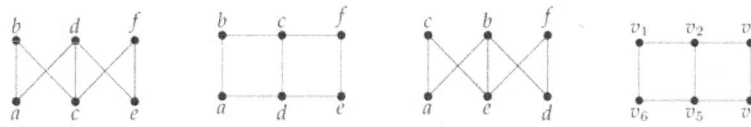

You might also think of a graph as a set of vertices and a set of edges, where each edge is a set of two vertices. Are the graphs below the same or different?

Graph 1:
$V = \{a, b, c, d, e\}$,
$E = \{\{a,b\}, \{a,c\}, \{a,d\}, \{a,e\}, \{b,c\}, \{d,e\}\}$.

Graph 2:
$V = \{v_1, v_2, v_3, v_4, v_5\}$,
$E = \{\{v_1,v_3\}, \{v_1,v_5\}, \{v_2,v_4\}, \{v_2,v_5\}, \{v_3,v_5\}, \{v_4,v_5\}\}$.

Before we start studying graphs, we need to agree upon what a graph is. While we almost always think of graphs as pictures (dots connected by lines) this is fairly ambiguous. Do the lines need to be straight? Does it matter how long the lines are or how large the dots are? Can there be two lines connecting the same pair of dots? Can one line connect three dots?

The way we avoid ambiguities in mathematics is to provide concrete and rigorous *definitions*. Crafting good definitions is not easy, but it is incredibly important. The definition is the agreed upon starting point from which all truths in mathematics proceed. Is there a graph with no edges? We have to look at the definition to see if this is possible.

We want our definition to be precise and unambiguous, but it also must agree with our intuition for the objects we are studying. It needs to be useful: we *could* define a graph to be a six legged mammal, but that would not let us solve any problems about bridges.

Activity 278

Write a definition for a *graph*. Make sure that your definition can be helpful in making sense of Activity 277.

Writing definitions is hard. You should compare what you came up with to the following:

> **Definition 3.1.1**
> A **graph** is an ordered pair $G = (V, E)$ consisting of a nonempty set V (called the **vertices**) and a set E (called the **edges**) of two-element subsets of V.

Strange. Nowhere in the definition is there talk of dots or lines. From the definition, a graph could be

$$(\{a,b,c,d\},\{\{a,b\},\{a,c\},\{b,c\},\{b,d\},\{c,d\}\}).$$

Here we have a graph with four vertices (the letters a, b, c, d) and four edges (the pairs $\{a,b\}, \{a,c\}, \{b,c\}, \{b,d\}, \{c,d\}$)).

Looking at sets and sets of 2-element sets is difficult to process. That is why we often draw a representation of these sets. We put a dot down for each vertex, and connect two dots with a line precisely when those two vertices are one of the 2-element subsets in our set of edges. Thus one way to draw the graph described above is this:

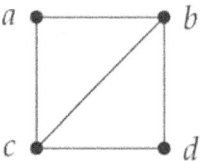

However we could also have drawn the graph differently. For example either of these:

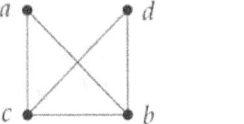   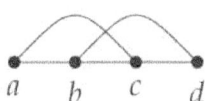

We should be careful about what it means for two graphs to be "the same." Actually, given our definition, this is easy: Are the vertex sets equal? Are the edge sets equal? We know what it means for sets to be equal, and graphs are nothing but a pair of two special sorts of sets.

> **Example 3.1.2**
> Are the graphs below equal?
>
> $G_1 = (\{a,b,c\},\{\{a,b\},\{b,c\}\}); \qquad G_2 = (\{a,b,c\},\{\{a,c\},\{c,b\}\})$
>
> equal?
>
> **Solution.** No. Here the vertex sets of each graph are equal, which is a good start. Also, both graphs have two edges. In the first graph, we have edges $\{a,b\}$ and $\{b,c\}$, while in the second graph we have edges $\{a,c\}$ and $\{c,b\}$. Now we do have $\{b,c\} = \{c,b\}$, so that is not the problem. The issue is that $\{a,b\} \neq \{a,c\}$. Since the edge sets of the two graphs are not equal (as sets), the graphs are not equal (as graphs).

Even if two graphs are not *equal*, they might be *basically* the same. The graphs in the previous example could be drawn like this:

Graphs that are basically the same (but perhaps not equal) are called **isomorphic**. We will give a precise definition of this term after a quick example:

Example 3.1.3

Consider the graphs:

$G_1 = \{V_1, E_1\}$ where $V_1 = \{a, b, c\}$ and $E_1 = \{\{a,b\}, \{a,c\}, \{b,c\}\}$;

$G_2 = \{V_2, E_2\}$ where $V_2 = \{u, v, w\}$ and $E_2 = \{\{u,v\}, \{u,w\}, \{v,w\}\}$.

Are these graphs the same?

Solution. The two graphs are NOT equal. It is enough to notice that $V_1 \ne V_2$ since $a \in V_1$ but $a \notin V_2$. However, both of these graphs consist of three vertices with edges connecting every pair of vertices. We can draw them as follows:

Clearly we want to say these graphs are basically the same, so while they are not equal, they will be *isomorphic*. The reason is we can rename the vertices of one graph and get the second graph as the result.

Intuitively, graphs are **isomorphic** if they are basically the same, or better yet, if they are the same except for the names of the vertices. To make the concept of renaming vertices precise, we give the following definitions:

Definition 3.1.4

An **isomorphism** between two graphs G_1 and G_2 is a bijection $f : V_1 \to V_2$ between the vertices of the graphs such that $\{a, b\}$ is an edge in G_1 if and only if $\{f(a), f(b)\}$ is an edge in G_2.

Two graphs are **isomorphic** if there is an isomorphism between them. In this case we write $G_1 \cong G_2$.

An isomorphism is simply a function which renames the vertices. It must be a bijection so every vertex gets a new name. These newly named vertices must be connected by edges precisely if they were connected by edges with their old names.

Example 3.1.5

Decide whether the graphs $G_1 = \{V_1, E_1\}$ and $G_2 = \{V_2, E_2\}$ are equal or isomorphic.

$V_1 = \{a, b, c, d\}, E_1 = \{\{a,b\}, \{a,c\}, \{a,d\}, \{c,d\}\}$
$V_2 = \{a, b, c, d\}, E_2 = \{\{a,b\}, \{a,c\}, \{b,c\}, \{c,d\}\}$

Solution. The graphs are NOT equal, since $\{a,d\} \in E_1$ but $\{a,d\} \notin E_2$. However, since both graphs contain the same number of vertices and same number of edges, they *might* be isomorphic (this is not enough in most cases, but it is a good start).

We can try to build an isomorphism. How about we say $f(a) = b$, $f(b) = c$, $f(c) = d$ and $f(d) = a$. This is definitely a bijection, but to make sure that the function is an isomorphism, we must make sure it *respects the edge relation*. In G_1, vertices a and b are connected by an edge. In G_2, $f(a) = b$ and $f(b) = c$ are connected by an edge. So far, so good, but we must check the other three edges. The edge $\{a, c\}$ in G_1 corresponds to $\{f(a), f(c)\} = \{b, d\}$, but here we have a problem. There is no edge between b and d in G_2. Thus f is NOT an isomorphism.

Not all hope is lost, however. Just because f is not an isomorphism does not mean that there is no isomorphism at all. We can try again. At this point it might be helpful to draw the graphs to see how they should match up.

Alternatively, notice that in G_1, the vertex a is adjacent to every other vertex. In G_2, there is also a vertex with this property: c. So build the bijection $g : V_1 \to V_2$ by defining $g(a) = c$ to start with. Next, where should we send b? In G_1, the vertex b is only adjacent to vertex a. There is exactly one vertex like this in G_2, namely d. So let $g(b) = d$. As for the last two, in this example, we have a free choice: let $g(c) = b$ and $g(d) = a$ (switching these would be fine as well).

We should check that this really is an isomorphism. It is definitely a bijection. We must make sure that the edges are respected. The four edges in G_1 are

$$\{a,b\}, \{a,c\}, \{a,d\}, \{c,d\}$$

Under the proposed isomorphism these become

$$\{g(a), g(b)\}, \{g(a), g(c)\}, \{g(a), g(d)\}, \{g(c), g(d)\}$$

$$\{c,d\}, \{c,b\}, \{c,a\}, \{b,a\}$$

which are precisely the edges in G_2. Thus g is an isomorphism, so $G_1 \cong G_2$

Sometimes we will talk about a graph with a special name (like K_n or the *Peterson graph*) or perhaps draw a graph without any labels. In this case we are really referring to *all* graphs isomorphic to any copy of that particular graph. A collection of isomorphic graphs is often called an **isomorphism class**.[1]

Activity 279
Consider the following two graphs:

G_1 $V_1 = \{a, b, c, d, e, f, g\}$
$E_1 = \{\{a,b\}, \{a,d\}, \{b,c\}, \{b,d\}, \{b,e\}, \{b,f\}, \{c,g\}, \{d,e\},$
$\{e,f\}, \{f,g\}\}$.

G_2 $V_2 = \{v_1, v_2, v_3, v_4, v_5, v_6, v_7\}$,
$E_2 = \{\{v_1, v_4\}, \{v_1, v_5\}, \{v_1, v_7\}, \{v_2, v_3\}, \{v_2, v_6\},$
$\{v_3, v_5\}, \{v_3, v_7\}, \{v_4, v_5\}, \{v_5, v_6\}, \{v_5, v_7\}\}$

(a) Let $f : G_1 \to G_2$ be a function that takes the vertices of Graph 1 to vertices of Graph 2. The function is given by the following table:

x	a	b	c	d	e	f	g
$f(x)$	v_4	v_5	v_1	v_6	v_2	v_3	v_7

Does f define an isomorphism between Graph 1 and Graph 2?

(b) Define a new function g (with $g \neq f$) that defines an isomorphism between Graph 1 and Graph 2.

(c) Is the graph pictured below isomorphic to Graph 1 and Graph 2? Explain.

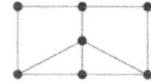

Activity 280
Is it possible for two *different* (non-isomorphic) graphs to have the same number of vertices and the same number of edges? What if the degrees of the vertices in the two graphs are the same (so both graphs have vertices with degrees 1, 2, 2, 3, and 4, for example)? Draw two such graphs or explain why not.

There are other relationships between graphs that we care about, other than equality and being isomorphic. For example, compare the following pair of graphs:

---

[1]This is not unlike geometry, where we might have more than one copy of a particular triangle. There instead of *isomorphic* we say *congruent*.

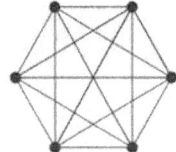 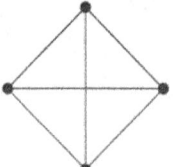

These are definitely not isomorphic, but notice that the graph on the right looks like it might be part of the graph on the left, especially if we draw it like this:

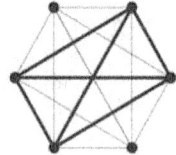

We would like to say that the smaller graph is a *subgraph* of the larger.

We should give a careful definition of this. In fact, there are two reasonable notions for what a subgroup should mean.

> **Definition 3.1.6**
> We say that $G' = (V', E')$ is a **subgraph** of $G = (V, E)$, and write $G' \subseteq G$, provided $V' \subseteq V$ and $E' \subseteq E$.
>
> We say that $G' = (V', E')$ is an **induced subgraph** of $G = (V, E)$ provided $V' \subseteq V$ and every edge in E whose vertices are still in V' is also an edge in E'.

Notice that every induced subgraph is also an ordinary subgraph, but not conversely. Think of a subgraph as the result of deleting some vertices and edges from the larger graph. For the subgraph to be an induced subgraph, we can still delete vertices, but now we only delete those edges that included the deleted vertices.

> **Example 3.1.7**
> Consider the graphs:
>
>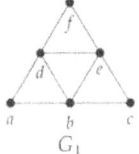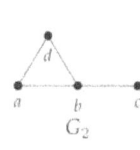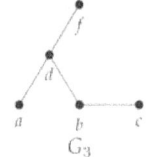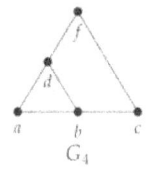
>
> $\qquad G_1 \qquad\qquad G_2 \qquad\qquad G_3 \qquad\qquad G_4$
>
> Here both G_2 and G_3 are subgraphs of G_1. But only G_2 is an *induced* subgraph. Every edge in G_1 that connects vertices in G_2 is also an edge in G_2. In G_3, the edge $\{a, b\}$ is in E_1 but not E_3, even though vertices a and b are in V_3.
>
> The graph G_4 is NOT a subgraph of G_1, even though it looks like all we did is remove vertex e. The reason is that in E_4 we have the edge $\{c, f\}$ but this is not an element of E_1, so we don't have the required $E_4 \subseteq E_1$.

Back to some basic graph theory definitions. Notice that all the graphs we have drawn above have the property that no pair of vertices is connected more than once, and no vertex is connected to itself. Graphs like these are sometimes called **simple**, although we will just call them *graphs*. This is because our definition for a graph says that the edges form a set of 2-element subsets of the vertices. Remember that it doesn't make sense to say a set contains an element more than once. So no pair of vertices can be connected by an edge more than once. Also, since each edge must be a set containing two vertices, we cannot have a single vertex connected to itself by an edge.

That said, there are times we want to consider double (or more) edges and single edge loops. For example, the "graph" we drew for the Bridges of Königsberg problem had double edges because there really are two bridges connecting a particular island to the near shore. We will call these objects **multigraphs**. This is a good name: a *multiset* is a set in which we are allowed to include a single element multiple times.

The graphs above are also **connected**: you can get from any vertex to any other vertex by following some path of edges. A graph that is not connected can be thought of as two separate graphs drawn close together. For example, the following graph is NOT connected because there is no path from a to b:

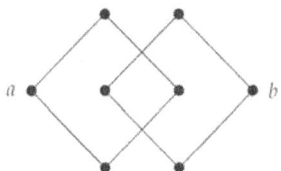

Most of the time, it makes sense to treat non-connected graphs as separate graphs (think of the above graph as two squares), so unless otherwise stated, we will assume all our graphs are connected.

Vertices in a graph do not always have edges between them. If we add all possible edges, then the resulting graph is called **complete**. That is, a graph is complete if every pair of vertices is connected by an edge. Since a graph is determined completely by which vertices are adjacent to which other vertices, there is only one complete graph with a given number of vertices. We give these a special name: K_n is the complete graph on n vertices.

Each vertex in K_n is adjacent to $n - 1$ other vertices. We call the number of edges emanating from a given vertex the **degree** of that vertex. So every vertex in K_n has degree $n - 1$. How many edges does K_n have? One might think the answer should be $n(n-1)$, since we count $n - 1$ edges n times (once for each vertex). However, each edge is incident to 2 vertices, so we counted every edge exactly twice. Thus there are $n(n-1)/2$ edges in K_n.

In general, if we know the degrees of all the vertices in a graph, we can find the number of edges. The sum of the degrees of all vertices will always be *twice* the number of edges, since each edge adds to the degree of two vertices. Notice this means that the sum of the degrees of all vertices in any graph must be even!

Example 3.1.8

At a recent math seminar, 9 mathematicians greeted each other by shaking hands. Is it possible that each mathematician shook hands with exactly 7 people at the seminar?

Solution. It seems like this should be possible. Each mathematician chooses one person to not shake hands with. But this cannot happen. We are asking whether a graph with 9 vertices can have each vertex have degree 7. If such a graph existed, the sum of the degrees of the vertices would be $9 \cdot 7 = 63$. This would be twice the number of edges (handshakes) resulting in a graph with 31.5 edges. That is impossible. Thus at least one (in fact an odd number) of the mathematicians must have shaken hands with an *even* number of people at the seminar.

One final definition: we say a graph is **bipartite** if the vertices can be divided into two sets, A and B, with no two vertices in A adjacent and no two vertices in B adjacent. The vertices in A can be adjacent to some or all of the vertices in B. If each vertex in A is adjacent to all the vertices in B, then the graph is a **complete bipartite graph**, and gets a special name: $K_{m,n}$, where $|A| = m$ and $|B| = n$. The graph in the houses and utilities puzzle is $K_{3,3}$.

Named Graphs.

Some graphs are used more than others, and get special names.

K_n The complete graph on n vertices.

$K_{m,n}$ The complete bipartite graph with sets of m and n vertices.

C_n The cycle on n vertices, just one big loop.

P_n The path on $n + 1$ vertices (so n edges), just one long path.

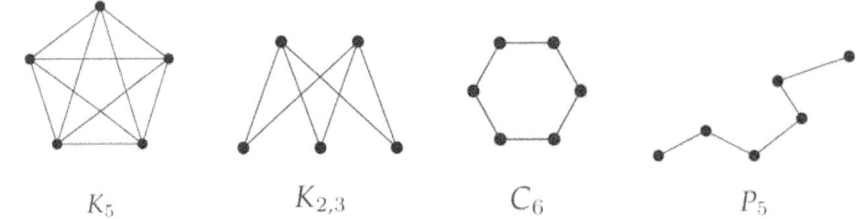

K_5 $K_{2,3}$ C_6 P_5

There are a lot of definitions to keep track of in graph theory. Here is a glossary of the terms we have already used and will soon encounter.

Graph Theory Definitions.

Graph

A collection of **vertices**, some of which are connected by **edges**. More precisely, a pair of sets V and E where V is a set of vertices and E is a set of 2-element subsets of V.

Adjacent
> Two vertices are **adjacent** if they are connected by an edge. Two edges are **adjacent** if they share a vertex.

Bipartite graph
> A graph for which it is possible to divide the vertices into two disjoint sets such that there are no edges between any two vertices in the same set.

Complete bipartite graph
> A bipartite graph for which every vertex in the first set is adjacent to every vertex in the second set.

Complete graph
> A graph in which every pair of vertices is adjacent.

Connected
> A graph is **connected** if there is a path from any vertex to any other vertex.

Chromatic number
> The minimum number of colors required in a proper vertex coloring of the graph.

Cycle
> A path (see below) that starts and stops at the same vertex, but contains no other repeated vertices.

Degree of a vertex
> The number of edges incident to a vertex.

Euler path
> A walk which uses each edge exactly once.

Euler circuit
> An Euler path which starts and stops at the same vertex.

Multigraph
> A **multigraph** is just like a graph but can contain multiple edges between two vertices as well as single edge loops (that is an edge from a vertex to itself).

Path A **path** is a walk that doesn't repeat any vertices (or edges) except perhaps the first and last. If a path starts and ends at the same vertex, it is called a **cycle**.

Planar
> A graph which can be drawn (in the plane) without any edges crossing.

Subgraph
> We say that H is a **subgraph** of G if every vertex and edge of H is also a vertex or edge of G. We say H is an **induced** subgraph of G if every vertex of H is a vertex of G and each pair of vertices in H are adjacent in H if and only if they are adjacent in G.

Tree A (connected) graph with no cycles. (A non-connected graph with no cycles is called a **forest**.) The vertices in a tree with degree 1 are called **leaves**.

Vertex coloring
An assignment of colors to each of the vertices of a graph. A vertex coloring is **proper** if adjacent vertices are always colored differently.

Walk A sequence of vertices such that consecutive vertices (in the sequence) are adjacent (in the graph). A walk in which no edge is repeated is called a **trail**, and a trail in which no vertex is repeated (except possibly the first and last) is called a **path**.

3.2 Walking Around Graphs

Puzzle 281

On the table rest 8 dominoes, as shown below. If you were to line them up in a single row, so that any two sides touching had matching numbers, what would the sum of the two end numbers be?

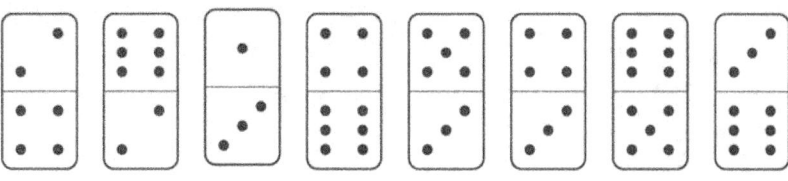

How might you use graph theory to solve the puzzle above? One thing to try would be to create a graph consisting of six vertices, one for each value that appears on the dominoes. Then connect the vertices if their numbers belong to the same domino. This seems reasonable, but then how would you represent the line of matched up dominoes the puzzle requests?

Say you start with domino $(4,6)$. That is an edge in our graph. Next we need to place a domino with a 6 on it (or a 4, but let's say the 4 will be an endpoint). This amounts to selecting another edge, one that is incident to the vertex 6. Say we pick $(6,5)$. The edge we must select after that will need to be incident to vertex 5. And so on.

The edges we select in the graph need to be linked up. It makes sense to think about the selection of edges as traveling through the graph, moving along edges from vertex to vertex. Let's capture this notion carefully.

Definition 3.2.1

A **walk** in a graph $G = (V, E)$ is a sequence of vertices $v_0, v_1, v_2, \ldots, v_n$ such that $\{v_i, v_{i+1}\} \in E$ for all $0 \leq i < n$. That is, consecutive vertices in the sequence are adjacent in the graph.

A **trail** is a walk that does not repeat any edges. If the trail starts and ends at the same vertex (i.e., $v_1 = v_n$) then it is called a **circuit**.

A **path** is a trail that does not repeat any vertices, except perhaps for $v_0 = v_n$. In the case that $v_0 = v_n$, the path is called a **cycle**.

Do these definitions capture what a walk/trail/path should mean in a graph? All three of these describe ways in which to move around in the graph, perhaps tracing along the edges without lifting up your pencil.

Note: some sources give slightly different definitions for these concepts. For example, sometimes a walk is defined as a alternating sequence of vertices and edges $v_1 e_1 v_2 e_2 v_3 \cdots e_{n-1} v_n$ with the requirement that v_i is incident to e_i, which is incident to v_{i+1}. We have not named our edges, so this definition would not make sense. Another common way to define *path* is as a subgraph isomorphic to P_n for some n. This is almost equivalent to our definition, except that we allow our paths to possibly be isomorphic to C_n for some n as well. Some sources use *path* where we use *walk* and use *simple path* where we use *path*.

Walks and paths are useful for considering all sorts of graph theoretic concepts. For example, you might want to find the **distance** between two vertices. This would be defined as the shortest path starting at one and ending at the other (where the **length** of a path is the number of edges in it). If you have a notion of distance, you can define the *center* and *radius* of a graph. Graphs are *connected* provided between any two vertices, there is a path (or equivalently, a walk).

3.2.1 Euler paths and circuits

To solve the domino problem, as well as the Bridges of Königsberg, we need to consider walks that use every edge exactly once.

> **Definition 3.2.2**
> A walk in a graph that uses every edge exactly once is called an **Euler path**. An Euler path that starts and ends at the same vertex is called an **Euler circuit**.

Although it is tradition, *Euler path* is a bad name. These should be called an **Euler trail** or **Euler walk**, since by our definitions, an Euler path is usually not a path at all. An Euler circuit is also sometimes called an **Euler tour**.

> **Activity 282**
> Our goal is to find a quick way to check whether a graph (or multigraph) has an Euler path or circuit.
>
> (a) Which of the graphs/multigraphs below have Euler paths? Which have Euler circuits?
>
>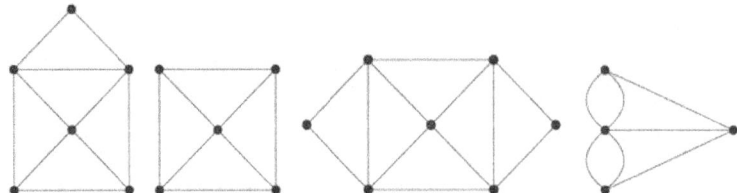
>
> (b) List the degrees of each vertex of the (multi)graphs above. Is there a connection between degrees and the existence of Euler paths and circuits?
>
> (c) Is it possible for a graph with a degree 1 vertex to have an Euler circuit? If so, draw one. If not, explain why not. What about an Euler path?
>
> (d) What if every vertex of the graph has degree 2. Is there an Euler path? An Euler circuit? Draw some graphs.
>
> (e) Below is *part* of a graph. Even though you can only see some of the vertices, can you deduce whether the graph will have an Euler path or circuit?

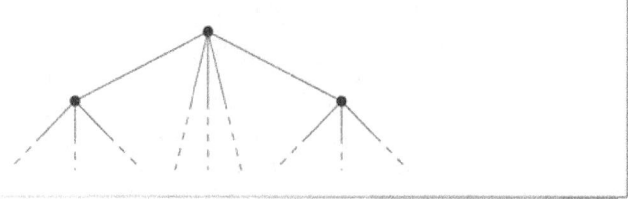

So what does it take for a graph to have an Euler path or circuit? Of course if a graph is not connected, there is no hope of finding such a path or circuit. For the rest of this section, assume all the graphs discussed are connected.

The bridges of Königsberg problem is really a question about the existence of Euler paths. There will be a route that crosses every bridge exactly once if and only if the multigraph below has an Euler path:

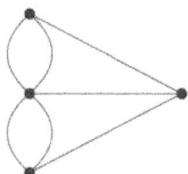

This graph is small enough that we could actually check every possible walk that does not reuse edges, and in doing so convince ourselves that there is no Euler path (let alone an Euler circuit). On small graphs which do have an Euler path, it is usually not difficult to find one. Our goal is to find a quick way to check whether a graph has an Euler path or circuit, even if the graph is quite large.

One way to guarantee that a graph does *not* have an Euler circuit is to include a "spike," a vertex of degree 1.

The vertex a has degree 1, and if you try to make an Euler circuit, you see that you will get stuck at the vertex. It is a dead end. That is, unless you start there. But then there is no way to return, so there is no hope of finding an Euler circuit. There is however an Euler path. It starts at the vertex a, then loops around the triangle. You will end at the vertex of degree 3.

You run into a similar problem whenever you have a vertex of any odd degree. If you start at such a vertex, you will not be able to end there (after traversing every edge exactly once). After using one edge to leave the starting vertex, you will be left with an even number of edges emanating from the vertex. Half of these could be used for returning to the vertex, the other half for leaving. So you return, then leave. Return, then leave. The only way to use up all the edges is to use the last one by leaving the vertex. On the other hand, if you have a vertex with odd degree that you do not start a path at, then you will eventually get stuck at that vertex. The path will use pairs of edges incident to the vertex to arrive and leave again. Eventually all but one of these edges will be used up, leaving only an edge to arrive by, and none to leave again.

What all this says is that if a graph has an Euler path and two vertices with odd degree, then the Euler path must start at one of the odd degree vertices and end at the other. In such a situation, every other vertex *must* have an even degree since we need an equal number of edges to get to those vertices as to leave them. How could we have an Euler circuit? The graph could not have any odd degree vertex as an Euler path would have to start there or end there, but not both. Thus for a graph to have an Euler circuit, all vertices must have even degree.

So there is clearly a connection between the parity of vertex degrees and Euler paths and circuits. Here are two theorems the describe that connection.

Theorem 3.2.3
A connected graph has an Euler circuit if and only if the degree of every vertex is even.

Theorem 3.2.4
A connected graph has an Euler path if and only if there are at most two vertices with odd degree.

Note that these theorems answer the bridges of Königsberg problem. The graph has all four vertices with odd degree, there is no Euler path through the graph. Thus there is no way for the townspeople to cross every bridge exactly once.

3.2.2 Proofs for Euler Paths and Circuits

Let's try to prove Theorems 3.2.3 and Theorem 3.2.4. Actually, we really only need to prove one of these, and the other one should follow from it.

Activity 283
Suppose we have already proved Theorem 3.2.3, that a graph has an Euler circuit if and only if every vertex has even degree. Suppose now we want to prove Theorem 3.2.4, that a graph has an Euler path if and only if it has at most two odd degree vertices.

(a) Given a graph with an Euler path, how can you get a graph that definitely has an Euler circuit? How did this affect the number of odd degree vertices? [Hint]

(b) Given a graph with at most two odd degree vertices, build a graph that has only even degree vertices. If you have an Euler circuit of your new graph, what would that Euler circuit become if you looked at your original graph? [Hint]

(c) What do each of the above tasks demonstrate? That is, write a proof of Theorem 3.2.4 assuming Theorem 3.2.3. Make sure you specify how each "direction" (implication and converse) of the proof is established by the tasks above.

Good, so we only need to prove Theorem 3.2.3.

Activity 284

Which of the two implications will be easier to prove? State both implications and say specifically what it would take to prove each. Say what would be required both for doing a direct proof and a proof by contrapositive.

Activity 285

Suppose you had a graph and were given an Euler circuit for that graph. How might you represent that Euler circuit? If you listed each edge in order, where each edge was given as a pair of vertices, how many times would each vertex appear in the list? What have you just shown?

The other direction is harder. If you know every vertex has even degree, you must now show there is an Euler circuit. One thing you could do is to give a procedure for building one. But then you must prove that this procedure is correct. We will come back to this question when we look at mathematical induction.

3.2.3 Hamilton Paths

Suppose you wanted to tour Königsberg in such a way where you visit each land mass (the two islands and both banks) exactly once. This can be done. In graph theory terms, we are asking whether there is a path which visits every vertex exactly once. Such a path is called a **Hamilton path** (or **Hamiltonian path**). We could also consider **Hamilton cycles**, which are Hamilton paths which start and stop at the same vertex.

Example 3.2.5

Determine whether the graphs below have a Hamilton path.

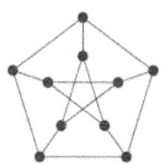 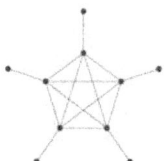

Solution. The graph on the left has a Hamilton path (many different ones, actually), as shown here:

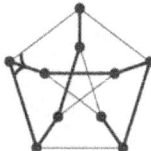

The graph on the right does not have a Hamilton path. You would need to visit each of the "outside" vertices, but as soon as you visit one, you get stuck. Note that this graph does not have an Euler path, although there are graphs with Euler paths but no Hamilton paths.

It appears that finding Hamilton paths would be easier because graphs often have more edges than vertices, so there are fewer requirements to be met. However, nobody knows whether this is true. There is no known simple test for whether a graph has a Hamilton path. For small graphs this is not a problem, but as the size of the graph grows, it gets harder and harder to check whether there is a Hamilton path. In fact, this is an example of a question which as far as we know is too difficult for computers to solve; it is an example of a problem which is NP-complete.

While a complete classification for which graphs have Hamilton paths is hard, there are some simpler things we can say. The following is one example.

Activity 286
Consider the following graph:

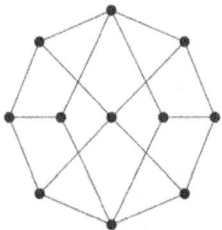

(a) Find a Hamilton path. Can your path be extended to a Hamilton cycle?

(b) Is the graph bipartite? If so, how many vertices are in each "part"?

(c) Use your answer to part (b) to prove that the graph has no Hamilton cycle.

(d) Suppose you have a bipartite graph G in which one part has at least two more vertices than the other. Prove that G does not have a Hamilton path.

3.3 Planar Graphs and Euler's Formula

Activity 287

When a connected graph can be drawn without any edges crossing, it is called **planar**. When a planar graph is drawn in this way, it divides the plane into regions called **faces**.

(a) Draw, if possible, two different planar graphs with the same number of vertices, edges, and faces.

(b) Draw, if possible, two different planar graphs with the same number of vertices and edges, but a different number of faces.

When is it possible to draw a graph so that none of the edges cross? If this *is* possible, we say the graph is **planar** (since you can draw it on the *plane*).

Notice that the definition of planar includes the phrase "it is possible to." This means that even if a graph does not look like it is planar, it still might be. Perhaps you can redraw it in a way in which no edges cross. For example, this is a planar graph:

That is because we can redraw it like this:

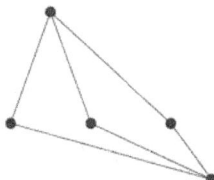

The graphs are the same, so if one is planar, the other must be too. However, the original drawing of the graph was not a **planar representation** of the graph.

When a planar graph is drawn without edges crossing, the edges and vertices of the graph divide the plane into regions. We will call each region a **face**. The graph above has 3 faces (yes, we *do* include the "outside" region as a face). The number of faces does not change no matter how you draw the graph (as long as you do so without the edges crossing), so it makes sense to ascribe the number of faces as a property of the planar graph.

WARNING: you can only count faces when the graph is drawn in a planar way. For example, consider these two representations of the same graph:

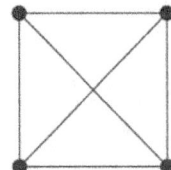
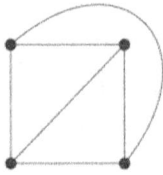

If you try to count faces using the graph on the left, you might say there are 5 faces (including the outside). But drawing the graph with a planar representation shows that in fact there are only 4 faces.

Is there a connection between the number of vertices (v), the number of edges (e) and the number of faces (f) in any connected planar graph? Let's investigate this. For each planar graph, we can associate a triple (v, e, f) containing the number of vertices, edges and faces. For example, the triple for K_4 drawn above is $(4, 6, 4)$.

Activity 288
Draw a lot of planar graphs and for each record the triple (v, e, f). What triples are possible? Is there a graph with the triple $(5, 9, 6)$? What about $(5, 7, 5)$?

If we want to understand the relationship between the number of vertices, edges and faces of a planar graph, we could first try to decide which triples can be realized by a graph and which cannot. At first this might seem like a daunting task. The key is to think of it recursively. Perhaps start with simple graphs and build up to more complicated ones.

Activity 289
What happens to the triple (v, e, f) when you add an edge to a graph?

(a) Start with a single pair of adjacent vertices (that is, the graph P_1). What is (v, e, f)? What happens to the triple if you extend the path to P_2? How do v, e, and f change?

(b) You could add another vertex and edge to P_2 (which would either give P_3 or a *star* graph), or you could add an edge connecting two vertices not already adjacent. What would these two operations do to (v, e, f)?

(c) Generalize. Say you know a specific triple (a, b, c) describes a graph G. Give two new triples that also describe some graphs, each with one more edge than G.

(d) Conjecture a relationship between v, e, and f that should hold for any connected planar graph.

It appears that whenever (v, e, f) describes some graph, then there is a graph with triple $(v + 1, e + 1, f)$ and a graph with triple $(v, e + 1, f + 1)$. Perhaps there are other ways we could add an edge we haven't thought of yet, but so far it seems like this completely describes how any triple must be created.

If this is the case, what could we conclude? Every time we add an edge, either the number of vertices or the number of faces increases by 1. Perhaps then $e = v + f$? But this is not true, since for P_1 we would have $1 = 2 + 1$, clearly false. Okay, at least we might guess $v + f - e$ is constant. And from P_1, we see that constant would need to be $2 + 1 - 1 = 2$. This leads us to conjecture the following theorem.

> **Theorem 3.3.1**
> For any planar graph with v vertices, e edges, and f faces, we have
> $$v - e + f = 2$$

We will soon see that this really is a theorem. The equation $v - e + f = 2$ is called **Euler's formula for planar graphs**. To prove this, we will want to somehow capture the idea of building up more complicated graphs from simpler ones. That is a job for mathematical induction!

3.3.1 Interlude: Mathematical Induction

The proof technique we will use to establish Euler's formula for planar graphs is called **mathematical induction**. This is a technique we use many times throughout our studies, so let's take a few moments to understand the logical structure for this style of proof and why it is a valid proof technique. Additional general practice on induction can be found in Section B.6.

> **Activity 290**
> What is the units digit of 6^n? How do you know?
>
> (a) Suppose you knew that the units digit of 6^{217} was a 2. Without calculating 6^{218} directly, can you say what its units digit would be?
>
> (b) Is the units digit of 6^{217} a 2? How do you know?

We often wish to prove a statement is true for all natural numbers n (or all integers larger than some smallest example). Mathematical induction uses the fact that the (infinitely many) cases we wish to prove are in a nice order with a least element. To show that 6^n has units digit 6 for all $n \geq 1$, we notice that $6^n = 6^{n-1} \cdot 6$. Assuming that we have already shown that 6^{n-1} has units digit 6, we can multiply this number by 6, and note that the units digit will still be a 6, since the units digit of $x \cdot 6$ is just 6 times the units digit of x, and $6 \cdot 6$ has units digit 6.

But this seems sneaky. Is it valid to assume that we already know the units digit of 6^{n-1}? That is the power of induction.

Suppose we hope to prove that $P(n)$ is true for all $n \geq 0$. What we will really prove is that $P(0)$ is true, and that for all k, if $P(k-1)$ is true, then $P(k)$ is true.

In other words, we are proving the implication

$$\forall k \, (P(k) \to P(k+1)).$$

This implication being true for all $k \geq 1$ really means we have proved infinitely many implications

$$P(0) \to P(1)$$
$$P(1) \to P(2)$$
$$P(2) \to P(3)$$
$$\vdots$$

Can we conclude $P(n)$ is true for all n? Well take any n. For example, $n = 3$. We know $P(0)$ is true, and the implication $P(0) \to P(1)$ is true, so we can conclude $P(1)$. From $P(1)$ and $P(1) \to P(2)$, we can conclude $P(2)$. From $P(2)$ and $P(2) \to P(3)$, we can conclude $P(3)$. If we wanted a different value of n, we would just need to keep going until we got up to it. Since every value of n can be reached by some finite number of steps starting at 0, we will know, for each n that $P(n)$ is true.

The above discussion is meant to explain why it is reasonable to believe that induction is a valid proof technique.[2] Once we believe induction works, we can just use it. That means just proving two things: $P(0)$ and for all $n \geq 0$, $P(n) \to P(n+1)$. We call $P(0)$ the **base case** and the implication the **inductive case**. Here is an example of what a proof by induction looks like.

Example 3.3.2

Prove by induction that for all $n \geq 0$, the number $n^2 + n$ is even.

Solution. First note that induction is NOT necessary here. We could give a non-inductive prove as follows.

Proof. Fix $n \geq 0$. Now $n^2 + n = n(n+1)$. Either n or $(n+1)$ is even, and the product of an even number and an odd number is even, so $n^2 + n$ is even. ∎

To contrast, here is a proof by induction.

Proof. First note that $0^2 + 0 = 0$ is even, so the case when $n = 0$ is true. Now assume for an arbitrary $n \geq 0$ that $n^2 + n$ is even. Consider the case for $n + 1$:

$$(n+1)^2 + n + 1 = n^2 + 2n + 1 + n + 1$$
$$= n^2 + n + 2n + 2.$$

But $n^2 + n$ is even, and $2n + 2$ is even, and the sum of two even numbers is even, so $(n+1)^2 + n + 1$ is even. ∎

Now that we have some sense for mathematical induction in general, let's consider how induction can be applied in graph theory.

---

[2]Technically, induction must be assumed as an axiom or deduced from the equivalent well-ordering principle which says that every subset of ℕ contains a least element; our justification for it is circular, since to prove the justification is valid would be done by induction itself.

First, we need something to play the role of n, since induction is used to prove statements are true for all natural numbers n. For the most part, we cannot think of all graphs as lined up in some natural order. But we could use induction *on* the number of edges of a graph (or number of vertices, or any other notion of size). That is we can prove that for all $n \geq 0$, all graphs with n edges have ....

Notice that the thing we are proving for all n is itself a universally quantified statement. This means that our base case become, "for all graphs with 0 edges ..." (which for not-necessarily connected graphs is infinitely many graphs). To prove the inductive case we *get to* assume that for all graphs with n edges ..., but then we *must* prove that for all graphs with $n + 1$ edges ....

Because of this added complication, the usual way to approach the inductive case for graphs is to start with an arbitrary graph that has $n + 1$ edges. Then remove an edge to drop down to a graph with n edges, which we know something about (our inductive hypothesis). Then make sure that moving back to the original graph, we maintain our desired property.

Let's try a simple example.

Activity 291

Recall that a **tree** is a connected graph with no cycles. We wish to prove that every tree with $v = n$ vertices has $e = n - 1$ edges. (This is actually a special case of Euler's formula for planar graphs, as a tree will always be a planar graph with 1 face).

(a) First, prove a lemma (not using induction): every tree contains at least two vertices of degree 1. [Hint]

(b) We will give a proof by induction on the number of vertices in a tree. What should the base case be? Prove it. [Hint]

(c) Now for the inductive case. Start with an arbitrary tree T with n vertices and assume that *all* trees with $n - 1$ vertices have $n - 2$ edges (why is the the right thing to assume)? Prove that T has $n - 1$ edges. [Hint]

We will use induction for many graph theory proofs, as well as proofs outside of graph theory. As our first example, we will prove Theorem 3.3.1

3.3.2 Proof of Euler's formula for planar graphs.

The proof we will give will be by induction on the number of edges of a graph. This means we will need a way to reduce the number of edges, so we can get down to our inductive hypothesis. There are two ways to remove an edge that will be useful here and for future problems.

A **deletion** is the result of removing an edge from the edge set (everything else in the graph stays the same). For a specific edge e in the graph G, we write the new graph as $G - e$.

Slightly more complicated is a **contraction**. The idea is we take an edge and shrink it down to a vertex. The two endpoints of the vertex collide and all their incident edges follow along and become incident to this new vertex consisting of the edge and two vertices contracted together. The resulting graph is written G/e (read "G contract e").

To be a little more precise, given a graph G and edge $e = \{u,v\}$, the graph G/e is formed by

1. Removing vertices u and v and all their incident vertices.

2. Adding a new vertex E and edges from E to each vertex that was previously adjacent to u or v.

Notice that G/e might be a multigraph (if u and v were both adjacent to a common vertex). If G was already a multigraph, and had two edges connecting u and v, then E will get a loop (since only one of the edges was contracted). We can avoid multigraphs by being careful which edge we contract along.

Activity 292
Let G be any planar graph with v vertices, e edges, and f faces.

(a) Let e_0 be any edge part of a cycle in G. Would it make more sense to remove e_0 using a deletion or a contraction? What will happen to $v - e + f$?

(b) Let e_0 be an edge incident to a vertex of degree 1. Should we remove e_0 by deletion or contraction? What will happen to $v - e + f$

Activity 293
Prove Euler's formula for planar graphs using induction on the number of edges in the graph. [Hint]

It is a good exercise to give proofs by induction on the number of vertices or the number of faces. You will need to think about how to remove a vertex or a face, and how doing so changes the other quantities.

3.4 Applications of Euler's Formula

We will now consider some applications of Euler's formula for planar graphs to graphs that are not necessarily planar.

3.4.1 Non-planar Graphs

Activity 294

For the complete graphs K_n, we would like to be able to say something about the number of vertices, edges, and (if the graph is planar) faces.

(a) How many vertices does K_3 have? How many edges? If K_3 is planar, how many faces should it have?

(b) How many vertices, edges and (if planar) faces would K_4, K_5 and $K_{2,3}$ each have?

(c) What about complete bipartite graphs? How many vertices, edges, and faces (if it were planar) does $K_{7,4}$ have?

Not all graphs are planar. If there are too many edges and too few vertices, then some of the edges will need to intersect. For example, consider K_5.

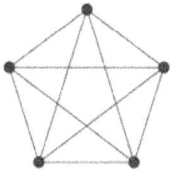

If you try to redraw this without edges crossing, you quickly get into trouble. There seems to be one edge too many. How could we prove this?

Activity 295

(a) The graph K_5 has 5 vertices and 10 edges. Explain why it would need to have 7 faces if it were planar.

(b) Now get at the number of faces another way: each face must be bordered by at least three edges. Why? Explain why we can conclude that $3f \leq 2e$. Where does the 2 come from?

(c) We now have that for K_5, the number of faces is $f = 7$, and also $3f \leq 20$. How is this possible? What can we conclude?

Before proceeding, consider carefully the style of proof used here. This is a **proof by contradiction**. We assumed the opposite of what we wanted to show. From that, we arrived at a contradiction, a statement that is necessarily false. Is it valid to conclude that our original assumption is false?

Recall the truth table for an implication $P \to Q$:

P	Q	$P \to Q$
T	T	T
T	F	F
F	T	T
F	F	T

In a proof by contradiction, we assume P (which will be the negation of our desired conclusion) and derive Q, a contradiction. That valid derivation allows us to conclude $P \to Q$ is *true*! But we know that Q is false. What does that say about P? If the implication is true and the consequent ("then" part) is false, it must be that we are in row 4 of the truth table. In that row, P is false. So we are justified in concluding $\neg P$.

In our case, P was the statement "K_5 is planar." So when we get a contradiction (a false Q), we can conclude that K_5 is not planar. Note that even though we are proving something about a graph that does not satisfy Euler's formula for planar graphs, by using a proof by contradiction, we get to *use* the formula.

Here are a few more examples of this proof strategy, specifically to show graphs are not planar.

Activity 296
The other simplest graph which is not planar is $K_{3,3}$

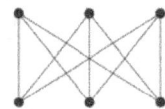

(a) Following the same proof outline as you used for K_5, what value do you find for f and what can you conclude from the inequality $3f \le 2e$? [Hint]

(b) The 3 in $3f \le 2e$ came from the observation that the any face in K_5 must be bounded by at least three edges. Is that the best we can do for a bipartite graph? Find and justify an improved inequality between f and e.

In general, if we let g be the size of the smallest cycle in a graph (g stands for *girth*, which is the technical term for this) then for any planar graph we have $gf \le 2e$. When this disagrees with Euler's formula, we know for sure that the graph cannot be planar.

You might wonder whether every graph that is not planar can be shown to be non-planar this way.

Activity 297
The graph show below is not planar. Let's prove it.

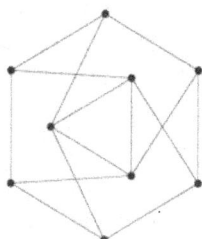

(a) What is the girth of this graph? What can you conclude from Euler's formula and the inequality $gf \leq 2e$?

(b) Try proving the following graph is not planar using our standard approach.

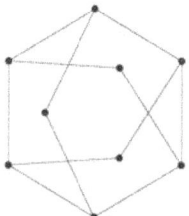

(c) Could it be that the original graph is planar but the subgraph is not? What can you conclude?

If a graph contains a non-planar subgraph, then the graph cannot itself be planar. This gives us another method of proving a given graph is not planar: find a non-planar graph, perhaps K_5 or $K_{3,3}$ inside it.

From the other direction, you can think of starting with K_5 or $K_{3,3}$ and adding edges as well as adding vertices in the middle of edges (called **subdividing** the edge) to build a larger graph that is definitely not planar. Surprisingly, *every* non-planar graph arises this way, a result called *Kuratowski's Theorem*.

Euler's formula can also be used to prove results about planar graphs.

Activity 298
Prove that any planar graph with v vertices and e edges satisfies $e \leq 3v-6$.

[Hint]

Activity 299
Prove that any planar graph must have a vertex of degree 5 or less. [Hint]

3.4.2 Polyhedra

Activity 300
A cube is an example of a convex polyhedron. It contains 6 identical squares for its faces, 8 vertices, and 12 edges. The cube is a **regular**

polyhedron (also known as a **Platonic solid**) because each face is an identical regular polygon and each vertex joins an equal number of faces.

There are exactly four other regular polyhedra: the tetrahedron, octahedron, dodecahedron, and icosahedron with 4, 8, 12 and 20 faces respectively. How many vertices and edges do each of these have?

Another area of mathematics that uses the terms "vertex," "edge," and "face" is geometry. A **polyhedron** is a geometric solid made up of flat polygonal faces joined at edges and vertices. We are especially interested in **convex** polyhedra, which means that any line segment connecting two points on the interior of the polyhedron must be entirely contained inside the polyhedron.[3]

Notice that since $8 - 12 + 6 = 2$, the vertices, edges and faces of a cube satisfy Euler's formula for planar graphs. This is not a coincidence. We can represent a cube as a planar graph by projecting the vertices and edges onto the plane. One such projection looks like this:

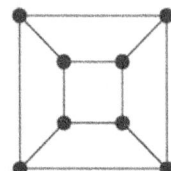

In fact, *every* convex polyhedron can be projected onto the plane without edges crossing. Think of placing the polyhedron inside a sphere, with a light at the center of the sphere. The edges and vertices of the polyhedron cast a shadow onto the interior of the sphere. You can then cut a hole in the sphere in the middle of one of the projected faces and "stretch" the sphere to lay down flat on the plane. The face that was punctured becomes the "outside" face of the planar graph.

The point is, we can apply what we know about graphs (in particular planar graphs) to convex polyhedra. Since every convex polyhedron can be represented as a planar graph, we see that Euler's formula for planar graphs holds for all convex polyhedra as well. We also can apply the same sort of reasoning we use for graphs in other contexts to convex polyhedra. For example, we know that there is no convex polyhedron with 11 vertices all of degree 3, as this would make 33/2 edges.

> **Example 3.4.1**
>
> Is there a convex polyhedron consisting of three triangles and six pentagons? What about three triangles, six pentagons and five heptagons (7-sided polygons)?
>
> **Solution.** How many edges would such polyhedra have? For the first proposed polyhedron, the triangles would contribute a total of 9 edges, and the pentagons would contribute 30. However, this counts each edge

---

[3]An alternative definition for convex is that the internal angle formed by any two faces must be less than 180 deg.

twice (as each edge borders exactly two faces), giving 39/2 edges, an impossibility. There is no such polyhedron.

The second polyhedron does not have this obstacle. The extra 35 edges contributed by the heptagons give a total of 74/2 = 37 edges. So far so good. Now how many vertices does this supposed polyhedron have? We can use Euler's formula. There are 14 faces, so we have $v - 37 + 14 = 2$ or equivalently $v = 25$. But now use the vertices to count the edges again. Each vertex must have degree *at least* three (that is, each vertex joins at least three faces since the interior angle of all the polygons must be less that $180°$), so the sum of the degrees of vertices is at least 75. Since the sum of the degrees must be exactly twice the number of edges, this says that there are strictly more than 37 edges. Again, there is no such polyhedron.

Activity 301
I'm thinking of a polyhedron containing 12 faces. Seven are triangles and four are quadrilaterals. The polyhedron has 11 vertices including those around the mystery face. How many sides does the last face have?

Activity 302
Consider some classic polyhedra.

(a) An *octahedron* is a regular polyhedron made up of 8 equilateral triangles (it sort of looks like two pyramids with their bases glued together). Draw a planar graph representation of an octahedron. How many vertices, edges and faces does an octahedron (and your graph) have?

(b) The traditional design of a soccer ball is in fact a (spherical projection of a) truncated icosahedron. This consists of 12 regular pentagons and 20 regular hexagons. No two pentagons are adjacent (so the edges of each pentagon are shared only by hexagons). How many vertices, edges, and faces does a truncated icosahedron have? Explain how you arrived at your answers. Bonus: draw the planar graph representation of the truncated icosahedron.

(c) Your "friend" claims that he has constructed a convex polyhedron out of 2 triangles, 2 squares, 6 pentagons and 5 octagons. Prove that your friend is lying. Hint: each vertex of a convex polyhedron must border at least three faces.

To conclude this application of planar graphs, consider the regular polyhedra. Above we claimed there are only five. How do we know this is true? We can prove it using graph theory.

Theorem 3.4.2
There are exactly five regular polyhedra.

Activity 303

Recall that a regular polyhedron has all of its faces identical regular polygons, and that each vertex has the same degree. Consider the cases, broken up by what the regular polygon might be.

(a) Case 1: Each face is a triangle. Let f be the number of faces and k the common degree of each vertex (since the polyhedron is regular). Find formulas for the number of edges and vertices in terms of f and k. Conclude from these the only possible values for f and k.

(b) Case 2: each face is a square. Again, consider the possible cases for f and k (and conclude there is only one: the cube).

(c) Case 3: Each face is a pentagon.

(d) Explain why it is not possible for each face to be a n-gon with $n \geq 6$.

---

[4]Notice that you can tile the plane with hexagons. This is an infinite planar graph; each vertex has degree 3. These infinitely many hexagons correspond to the limit as $f \to \infty$ to make $k = 3$.

3.5 Coloring

Puzzle 304

Mapmakers in the fictional land of Euleria have drawn the borders of the various dukedoms of the land. To make the map pretty, they wish to color each region. Adjacent regions must be colored differently, but it is perfectly fine to color two distant regions with the same color. What is the fewest colors the mapmakers can use and still accomplish this task?

Puzzle 305

How many colors do you need to color the faces of a cube so that no two faces that share an edge are colored the same? Is there a polyhedron that would require four colors to do this? Is there a polyhedron that requires five?

Perhaps the most famous graph theory problem is how to color maps.

> Given any map of countries, states, counties, etc., how many colors are needed to color each region on the map so that neighboring regions are colored differently?

Actual map makers usually use around seven colors. For one thing, they require watery regions to be a specific color, and with a lot of colors it is easier to find a permissible coloring. We want to know whether there is a smaller palette that will work for any map.

How is this related to graph theory? Well, if we place a vertex in the center of each region (say in the capital of each state) and then connect two vertices if their states share a border, we get a graph. Coloring regions on the map corresponds to coloring the vertices of the graph. Since neighboring regions cannot be colored the same, our graph cannot have vertices colored the same when those vertices are adjacent.

In general, given any graph G, a coloring of the vertices with k colors is called a k-**coloring**. If the coloring has the property that adjacent vertices are colored differently, then the coloring is called **proper**. Every graph has a proper k-coloring for some k. For example, you could color every vertex with a different color. But often you can do better. The smallest k for which there is a proper k-coloring is called the **chromatic number** of the graph, written $\chi(G)$.

> **Example 3.5.1**
> Find the chromatic number of the graphs below.
>
>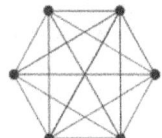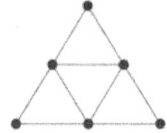
>
> **Solution.** The graph on the left is K_6. The only way to properly color the graph is to give every vertex a different color (since every vertex is adjacent to every other vertex). Thus the chromatic number is 6.
>
> The middle graph can be properly colored with just 3 colors (Red, Blue, and Green). For example:
>
>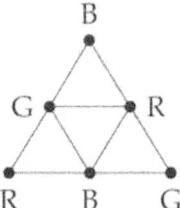
>
> There is no way to color it with just two colors, since there are three vertices mutually adjacent (i.e., a triangle). Thus the chromatic number is 3.
>
> The graph on the right is just $K_{2,3}$. As with all bipartite graphs, this graph has chromatic number 2: color the vertices on the top row red and the vertices on the bottom row blue.

It appears that there is no limit to how large chromatic numbers can get; the graph K_n has chromatic number n. So how could there possibly be an answer to the original map coloring question? If the chromatic number of graph can be arbitrarily large, then it seems like there would be no upper bound to the number of colors needed for any map. But there is.

The key observation is that while it is true that for any number n, there is a graph with chromatic number n, only some graphs arrive as representations of maps. If you convert a map to a graph, the edges between vertices correspond to borders between the countries. So you should be able to connect vertices in such a way where the edges do not cross. In other words, the graphs representing maps are all *planar*!

So the question is, what is the largest chromatic number of any planar graph? The answer is the infamous Four Color Theorem.

> **Theorem 3.5.2 The Four Color Theorem.**
> *If G is a planar graph, then the chromatic number of G is less than or equal to 4. Thus any map can be properly colored with 4 or fewer colors.*

We will not prove this theorem. Really. Even though the theorem is easy to state and understand, the proof is not. In fact, there is currently no "easy" known proof of the theorem. The current best proof still requires powerful computers to check an *unavoidable set* of 633 *reducible configurations*. The idea is that every graph must contain one of these reducible configurations (this fact also needs to be checked by a computer) and that reducible configurations can, in fact, be colored in 4 or fewer colors.

Of course if every planar graph has chromatic number at most 4, then every graph has chromatic number no more than 6. Proving this (even without using the 4 color theorem) is not very difficult, and this at least shows that the chromatic numbers for planar graphs is bounded.

Activity 306
Prove that every planar graph can be properly vertex colored with 6 or fewer colors. [Hint]

In 1879, Alfred Kempe published a proof of the Four Color Theorem. It wasn't until 11 years later that Percy John Heawood found an error in the proof. However, the failed proof did lead to a nice proof that every planar graph has chromatic number at most five. The next activity walks you through that proof.

Activity 307
We will prove the Five Color Theorem, that every planar graph has a proper vertex coloring using five (or fewer) colors. We proceed by induction on the number of vertices. So assume that G is a planar graph with n vertices, and all planar graphs with fewer vertices have a proper 5-coloring. We will use $\{1, 2, 3, 4, 5\}$ as our set of colors.

(a) Let v be a vertex of degree five or less (why can we do this?). Let H be the graph resulting from deleting v and all its incident edges. What can you say about H?

(b) Let v_1, v_2, v_3, v_4, v_5 be the vertices adjacent to v in G. Why is it okay to assume there really are five and further that these five vertices are colored distinctly from each other? [Hint]

(c) Assume $v_1, \ldots, v_5$ are drawn in the plane in clockwise order around v, and further that they are colored $1, 2, 3, 4, 5$ respectively. Now consider the induced subgraph $H_{1,3}$ of H consisting of all vertices colored 1 and 3 (and their edges). Why can we assume that there is a path from v_1 to v_3 contained in $H_{1,3}$? [Hint]

(d) Now consider $H_{2,4}$, the induced subgraph of H of vertices colored 2 and 4. Explain why their cannot be a path connecting v_2 and v_4, and why this completes the proof. [Hint]

You will notice that the proof above doesn't ever use v_5, so it is tempting to apply this same technique to prove the four color theorem. See if you can find the error in doing so in fewer than 11 years.

3.5.1 Coloring in General

There are plenty of reasons to color graphs other than cartography.

Activity 308
The math department plans to offer 10 classes next semester. Some classes cannot run at the same time (perhaps they are taught by the same professor, or are required for seniors).

Class:	Conflicts with:
A	D I
B	D I J
C	E F I
D	A B F
E	H I
F	I
G	J
H	E I J
I	A B C E F H
J	B G H

How many different time slots are needed to teach these classes (and which should be taught at the same time)? More importantly, how could we use graph coloring to answer this question?

Activity 309
Radio stations broadcast their signal at certain frequencies. However, there are a limited number of frequencies to choose from, so nationwide many stations use the same frequency. This works because the stations are far enough apart that their signals will not interfere; no one radio could pick them up at the same time.

Suppose 10 new radio stations are to be set up in a currently unpopulated (by radio stations) region. The radio stations that are close enough to each other to cause interference are recorded in the table below. What is the fewest number of frequencies the stations could use.

	KQEA	KQEB	KQEC	KQED	KQEE	KQEF	KQEG	KQEH	KQEI	KQEJ
KQEA			X			X	X			X
KQEB			X	X						
KQEC	X					X	X			X
KQED		X			X	X		X		
KQEE									X	
KQEF	X		X	X			X			X
KQEG	X		X			X				X
KQEH				X					X	
KQEI					X			X		X
KQEJ	X		X			X	X		X	

In the example above, the chromatic number was 5, but this is not a counterexample to the Four Color Theorem, since the graph representing the radio stations is not planar. It would be nice to have some quick way to find the chromatic number of a (possibly non-planar) graph. It turns out nobody knows whether an efficient algorithm for computing chromatic numbers exists.

While we might not be able to find the exact chromatic number of graph easily, we can often give a reasonable range for the chromatic number. In other words, we can give upper and lower bounds for chromatic number.

This is actually not very difficult: for every graph G, the chromatic number of G is at least 1 and at most the number of vertices of G.

What? You want *better* bounds on the chromatic number? Well, let's see what we can do.

Activity 310

(a) Find the chromatic number of the graph below. How do you know the chromatic number is not larger or smaller than your answer?

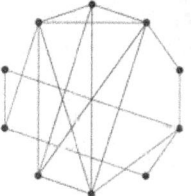

[Hint]

(b) For any graph G, if H is a subgraph with chromatic number k, what can you say about the chromatic number of G?

(c) Define the **clique number** of a graph to be the largest n for which the graph contains a copy of K_n as a subgraph (a **clique** in a graph is a set of vertices that are pairwise adjacent, so a copy of K_n for some n). State and prove a theorem relating the clique number and the chromatic number of a graph.

Could it be that the *only* way for a graph to have chromatic number n is for it to contain an n-clique?

Activity 311

(a) Find a graph that has chromatic number 3 even though it does not contain K_3 as a subgraph.

(b) Find a graph that has chromatic number 4 even though it does not contain K_4 as a subgraph. [Hint]

(c) Challenge: Find a graph that has chromatic number 4 even though it does not contain K_3 as a subgraph. [Hint]

Graphs that do not contain K_3 are called **triangle-free**. In 1955, Jan Mycielski gave a construction which when applied to any triangle-free graph with chromatic number n produced a triangle-free graph with chromatic number $n + 1$ (the resulting graph is called the **Mycielskian** of the original graph). This proves that the there are graphs without even K_3 that have arbitrarily high chromatic number, so the difference between the clique number and chromatic number can be arbitrarily large.

There are times when the chromatic number of G is *equal* to the clique number. These graphs have a special name; they are called **perfect** (actually for a graph to be perfect, every induced subgraph must have this property). If you know that a graph is perfect, then finding the chromatic number is simply a matter of searching for the largest clique.[4] However, not all graphs are perfect, and searching for the largest clique isn't always easy anyway.

What about an upper bound? For planar graphs, $\chi(G) \leq 4$. Also, if a graph has a proper k-coloring, then $\chi(G) \leq k$, but this requires attempting a coloring first. Can we do better? It is helpful to recall our attempts to prove the 6- and 5-color theorems.

Activity 312
Let G be any graph and let v be a vertex with degree k. Let $H = G - v$ be the subgraph resulting from removing v and all its incident edges.
Suppose $\chi(H) < k$. What can you conclude about $\chi(G)$? [Hint]

Some notation will come in handy. For a vertex v, write $d(v)$ for the degree of v (i.e., the number of vertices incident to it). We write $\Delta(G)$ for the **maximum degree of G**, which is $\Delta(G) = \max\{d(v) : v \in G\}$.

Activity 313
Prove that for any graph G, the chromatic number is no more than one more than the maximum degree. That is, $\chi(G) \leq \Delta(G) + 1$. [Hint]

---

[4]There are special classes of graphs which can be proved to be perfect. One such class is the set of **chordal** graphs, which have the property that every cycle in the graph contains a **chord**—an edge between two vertices of the cycle which are not adjacent in the cycle.

There are graphs for which $\chi(G) = \Delta(G) + 1$. For example, G might be an odd cycle, which has $\Delta(G) = 2$ but $\chi(G) = 3$. Also note that $\Delta(K_n) = n - 1$ but $\chi(K_n) = n$.

Surprisingly, for connected graphs, these are the only examples of this.

Theorem 3.5.3 Brooks' Theorem.
Any connected graph G satisfies $\chi(G) \le \Delta(G)$, unless G is a complete graph or an odd cycle, in which case $\chi(G) = \Delta(G) + 1$.

The proof of this theorem is a little tricky. We will be content with the following special case.

Activity 314
Prove by induction on vertices that any graph G which contains at least one vertex of degree less than $\Delta(G)$ (the maximal degree of all vertices in G) has chromatic number at most $\Delta(G)$.

3.5.2 Coloring Edges

The chromatic number of a graph tells us about coloring vertices, but we could also ask about coloring edges. Just like with vertex coloring, we might insist that edges that are adjacent must be colored differently. Here, we are thinking of two edges as being adjacent if they are incident to the same vertex. The least number of colors required to properly color the edges of a graph G is called the **chromatic index** of G, written $\chi'(G)$.

Activity 315
Six friends decide to spend the afternoon playing chess. Everyone will play everyone else once. They have plenty of chess sets but nobody wants to play more than one game at a time. Games will last an hour (thanks to their handy chess clocks). How many hours will the tournament last?

Interestingly, if one of the friends in the above example left, the remaining 5 chess-letes would still need 5 hours: the chromatic index of K_5 is also 5.

In general, what can we say about chromatic index? Certainly $\chi'(G) \ge \Delta(G)$. But how much higher could it be? Only a little higher.

Theorem 3.5.4 Vizing's Theorem.
For any graph G, the chromatic index $\chi'(G)$ is either $\Delta(G)$ or $\Delta(G) + 1$.

At first this theorem makes it seem like chromatic index might not be very interesting. However, deciding which case a graph is in is not always easy. Graphs for which $\chi'(G) = \Delta(G)$ are called *class 1*, while the others are called *class 2*. Bipartite graphs always satisfy $\chi'(G) = \Delta(G)$, so are class 1 (this was proved by König in 1916, decades before Vizing proved his theorem in 1964). In 1965 Vizing proved that all planar graphs with $\Delta(G) \ge 8$ are of class 1, but this does not hold for all planar graphs with $2 \le \Delta(G) \le 5$. Vizing conjectured that all planar graphs with $\Delta(G) = 6$ or $\Delta(G) = 7$ are class 1; the $\Delta(G) = 7$ case was proved in 2001 by Sanders and Zhao; the $\Delta(G) = 6$ case is still open.

3.5.3 Ramsey Theory

There is another interesting way we might consider coloring edges, quite different from what we have discussed so far. What if we colored every edge of a graph either red or blue. Can we do so without, say, creating a *monochromatic* triangle (i.e., an all red or all blue triangle)? Certainly for some graphs the answer is yes.

> **Activity 316**
> Find the largest number n so that there is a 2-coloring of the edges of K_n with no monochromatic triangle. Prove your answer is correct. This will involve proving that for *any* 2-coloring of the edges of K_{n+1}, there *is* a monochromatic triangle.

More generally, we can ask for the smallest number $R = R(m, n)$ such that no matter how we color the edges of K_R using red and blue, there will either be a red copy of K_m or a blue copy of K_n. The number $R(m, n)$ is called a **Ramsey number**. The previous activity proves that $R(3, 3) = 6$.

> **Activity 317**
> Show that among ten people, there are either four mutual acquaintances or three mutual strangers. What does this say about $R(4, 3)$?

> **Activity 318**
> Find a way to color the edges of K_8 with red and blue so that there is no red K_4 and no blue K_3. [Hint]

> **Activity 319**
> Find $R(4, 3)$. [Hint]

At this point you might wonder whether $R(m, n)$ even exists for all m and n. One way to prove this is to give a bound for it.

> **Activity 320**
> Prove that $R(m, n) \leq R(m - 1, n) + R(m, n - 1)$. [Hint]

> **Activity 321**
>
> (a) What does the equation in Activity 320 tell us about $R(4, 4)$?
>
> (b) Consider 17 people arranged in a circle such that each person is acquainted with the first, second, fourth, and eighth person to the

right and the first, second, fourth, and eighth person to the left. can you find a set of four mutual acquaintances? Can you find a set of four mutual strangers? [Hint]

(c) What is $R(4,4)$?

It turns out that $R(4,4) = 18$ and $R(5,4) = R(4,5) = 25$. Beyond that, not much is known. For example, $R(5,5)$ is at least 43 and at most 48. The gaps get larger as m and n increase: $R(10,10)$ is at least 798 and no more than 23556.

We can also color the edges with more colors. For example, $R(3,3,3) = 17$, which says that if you color the edges of K_{17} red, blue and green, there is guaranteed to be a monochromatic triangle in one of these colors.

3.6 Matching in Bipartite Graphs

In any **matching** is a subset M of the edges for which no two edges of M are incident to a common vertex. If every vertex belongs to exactly one of the edges, we say the matching is **perfect**. Our goal is to discover some criterion for when a bipartite graph has a prefect matching.

Suppose you have a bipartite graph G. This will consist of two sets of vertices A and B with some edges connecting some vertices of A to some vertices in B (but of course, no edges between two vertices both in A or both in B). A **matching** of G is a set of **independent edges**, meaning no two edges in the set are adjacent. If every vertex in G is incident to exactly one edge in the matching, we call the matching **perfect**. If a bipartite graph has a perfect matching, then $|A| = |B|$, but in general, we could have a **matching of** A, which will mean that every vertex in A is incident to an edge in the matching.

Some context might make this easier to understand. Think of the vertices in A as representing students in a class, and the vertices in B as representing presentation topics. We put an edge from a vertex $a \in A$ to a vertex $b \in B$ if student a would like to present on topic b. Of course, some students would want to present on more than one topic, so their vertex would have degree greater than 1. As the teacher, you want to assign each student their own unique topic. Thus you want to find a matching of A: you pick some subset of the edges so that each student gets matched up with exactly one topic, and no topic gets matched to two students.[5]

> **Activity 322**
> Does the graph below contain a perfect matching? If so, find one.
>
>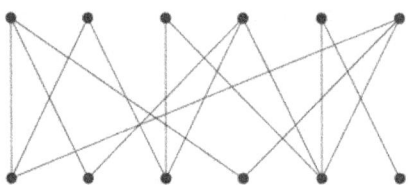

The question is: when does a bipartite graph contain a matching of A? To begin to answer this question, consider what could prevent the graph from containing a matching. This will not necessarily tell us a condition when the graph *does* have a matching, but at least it is a start.

> **Activity 323**
> Draw as many fundamentally different examples of bipartite graphs which do NOT have perfect matchings. Your goal is to find all the possible obstructions to a graph having a perfect matching. Write down the *necessary* conditions for a graph to have a matching (that is, fill in the

---

[5]The standard example for matchings used to be the *marriage problem* in which A consisted of the men in the town, B the women, and an edge represented a marriage that was agreeable to both parties. A matching then represented a way for the town elders to marry off everyone in the town, no polygamy allowed.

blank: If a graph has a matching, then _____). Then ask yourself whether these conditions are sufficient (is it true that if _____, then the graph has a matching?).

Activity 324

Find a matching of the bipartite graphs below or explain why no matching exists.

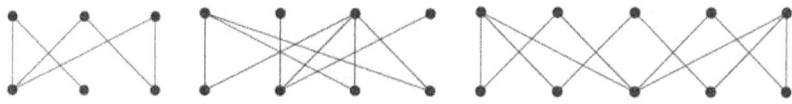

One way G could not have a matching is if there is a vertex in A not adjacent to any vertex in B (so having degree 0). What else? What if two students both like the same one topic, and no others? Then after assigning that one topic to the first student, there is nothing left for the second student to like, so it is very much as if the second student has degree 0. Or what if three students like only two topics between them. Again, after assigning one student a topic, we reduce this down to the previous case of two students liking only one topic. We can continue this way with more and more students.

It should be clear at this point that if there is every a group of n students who as a group like $n-1$ or fewer topics, then no matching is possible. This is true for any value of n, and any group of n students.

To make this more graph-theoretic, say you have a set $S \subseteq A$ of vertices. Define $N(S)$ to be the set of all the **neighbors** of vertices in S. That is, $N(S)$ contains all the vertices (in B) which are adjacent to at least one of the vertices in S. (In the student/topic graph, $N(S)$ is the set of topics liked by the students of S.) Our discussion above can be summarized as follows:

Lemma 3.6.1 Matching Condition.

If a bipartite graph $G = \{A, B\}$ has a matching of A, then

$$|N(S)| \geq |S|$$

for all $S \subseteq A$.

Activity 325

Write a careful proof of the matching condition above.

Is the converse true? Suppose G satisfies the matching condition $|N(S)| \geq |S|$ for all $S \subseteq A$ (every set of vertices has at least as many neighbors than vertices in the set). Does that mean that there is a matching? Surprisingly, yes. The obvious necessary condition is also sufficient.[6] This is a theorem first proved by Philip Hall in 1935.[7]

> **Theorem 3.6.2 Hall's Marriage Theorem.**
> *Let G be a bipartite graph with sets A and B. Then G has a matching of A if and only if*
> $$|N(S)| \geq |S|$$
> *for all $S \subseteq A$.*

There are a few different proofs for this theorem; we will consider one that gives us practice thinking about paths in graphs.

> **Activity 326**
> Every bipartite graph (with at least one edge) has a matching, even if it might not be perfect. Thus we can look for the largest matching in a graph. If that largest matching includes all the vertices, we have a perfect matching.
>
> Your "friend" claims that she has found the largest matching for the graph below (her matching is in bold). She explains that no other edge can be added, because all the edges not used in her partial matching are connected to matched vertices. Is she correct?
>
>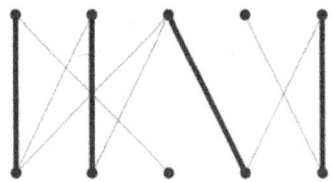

> **Activity 327**
> One way you might check to see whether a partial matching is maximal is to construct an **alternating path**. This is a path whose adjacent edges alternate between edges in the matching and edges not in the matching (no edge can be used more than once, since this is a path). If an alternating path starts and stops with vertices that are *not* matched, (that is, these vertices are not incident to any edge in the matching) then the path is called an **augmenting path**.
>
> (a) Find the largest possible alternating path for the matching of your friend's graph. Is it an augmenting path? How would this help you find a larger matching?

---

[6] This happens often in graph theory. If you can avoid the obvious counterexamples, you often get what you want.
[7] There is also an infinite version of the theorem which was proved by the unrelated Marshal Hall, Jr.

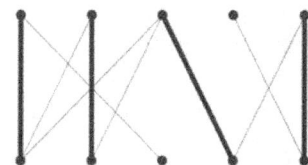

(b) Find the largest possible alternating path for the matching below. Are there any augmenting paths? Is the matching the largest one that exists in the graph?

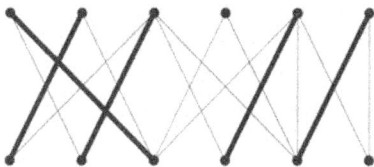

If a graph does not have a perfect matching, then any of its maximal matchings must leave a vertex unmatched. In particular, there cannot be an augmenting path starting at such a vertex (otherwise the maximal matching would not be maximal). Thus to prove Theorem 3.6.2, it would be sufficient to prove that the matching condition guarantees that every non-perfect matching has an augmenting path.

Activity 328
Let M be a matching of G that leaves a vertex $a \in A$ unmatched. We will find an augmenting path starting at a.

(a) Consider all the alternating paths starting at a and ending in A. Are any augmenting paths?

(b) Let A' be all the end vertices of alternating paths from above. Let $S = A' \cup \{a\}$. Explain why there must be some $b \in B$ that is adjacent to a vertex in S but not part of any of the alternating paths.

(c) Finish the proof. [Hint]

In addition to its application to marriage and student presentation topics, matchings have applications all over the place. We conclude with one such example.

Activity 329
Suppose you deal 52 regular playing cards into 13 piles of 4 cards each. Prove that you can always select one card from each pile to get one of each of the 13 card values Ace, 2, 3, ..., 10, Jack, Queen, and King. [Hint]

Activity 330

For many applications of matchings, it makes sense to use bipartite graphs. You might wonder, however, whether there is a way to find matchings in graphs in general.

(a) For which n does the complete graph K_n have a matching?

(b) Prove that if a graph has a matching, then $|V|$ is even.

(c) Is the converse true? That is, do all graphs with $|V|$ even have a matching?

(d) What if we also require the matching condition? Prove or disprove: If a graph with an even number of vertices satisfies $|N(S)| \geq |S|$ for all $S \subseteq V$, then the graph has a matching.

Appendix A
Group Projects

1. Paths and Binomial Coefficients
2. On Triangular Number
3. Fibonacci Numbers
4. Generalized Pascal Triangles
5. A Pascal-like Triangle of Eulerian Numbers
6. Leibniz Harmonic Triangle
7. The Twelve Days of Christmas

Project 1: Paths and Binomial Coefficients

1. A path consists of vertical and horizontal line segments of length 1. Determine the number of paths

   (a) of length 7 from the origin $(0,0)$ and $(4,3)$.

   (b) of length $a+b$ from $(0,0)$ and (a,b).

2. A lattice point in the plane is a point (m,n) with $m, n \in \mathbb{Z}$.

   (a) How many paths of length 4 are there from $(0,0)$ to lattice points on the line $y = -x + 4$?

   (b) Determine the number of paths of length 10 from $(0,0)$ to the lattice points on $y = 10 - x$.

3. Give a geometrical (path argument) for the identity $2^n = \binom{n}{0} + \binom{n}{1} + \ldots + \binom{n}{n}$ using the idea in Exercise A.2.

4. Generalize Exercise A.1 to three dimensions; introduce obstructions; use inclusion-exclusion.

5. Give a path proof for $\binom{n}{k} = \binom{n-1}{k} + \binom{n-1}{k-1}$.

6. How many lattice paths from $(0,0)$ to all lattice points on the line $x + 2y = n$ are there?

7. Determine the number of lattice paths from $(0,0,0)$ to lattice points on that portion of the plane $x + y + z = 2$ that lies in the first octant.

8. Generalize Exercise A.7.

9. Determine the number of lattice paths from $(0,0)$ to (n,n) that do not cross over (they may touch) the line $y = x$.

Project 2: On Triangular Numbers

Let $T_n = \frac{n(n+1)}{2}$ denote the nth triangular number.

1. Give a geometric interpretation for $T_n + T_{n+1} = (n+1)^2$.
2. Verify that $T_n + T_m + mn = T_{m+n}$ algebraically.
3. Give a geometric interpretation of the formula in Problem 2.
4. Verify that $T_n^2 + T_{n-1}^2 = T_{n^2}$.
5. Verify the following:

    (a) $3T_n + T_{n+1} = T_{2n+1}$

    (b) $3T_n + T_{n-1} = T_{2n}$ (also give a geometrical proof)

6. Determine the values $T_6, T_{66}, T_{666}, T_{6666}, \ldots$. Make a general statement and prove it.
7. Determine the formula for $T_3 + T_6 + T_9 + \ldots + T_{3n}$.
8. Verify:

    (a) $1 + 3 + 6 + 10 + 15 = 1^2 + 3^2 + 5^2$

    (b) $1 + 3 + 6 + 10 + 15 + 21 = 2^2 + 4^2 + 6^2$

    (c) Generalize and prove your generalization.

Project 3: Fibonacci

1. Briefly address history.
2. Using a variety of proof methods (geometric, induction, matrices, recursion, telescoping sum) prove identities such as:

   (a) $F_1 + F_2 + \ldots + F_n = F_{n+2} - 1$

   (b) $F_1^2 + F_2^2 + \ldots + F_n^2 = F_n F_{n+1}$

   (c) $F_{n+1}^2 - F_n F_{n+2} = (-1)^n$

   (d) $F_{n+2} = \binom{n+1}{0} + \binom{n}{1} + \binom{n-1}{2} + \cdots$

3. Prove the "Binet Formula," $F_n = \frac{\alpha^n - \beta^n}{\alpha - \beta}$, where α, β satisfy $x^2 - x - 1 = 0$.

   (a) By induction.

   (b) Using generating series.

4. Use the Binet Formula to prove,
$$\binom{n}{0}F_0 + \binom{n}{1}F_1 + \ldots + \binom{n}{n}F_0 = F_{2n}.$$

5. Investigate the geometric interpretations of the F_n.

6. Prove that,
$$F_n = \left[\binom{n}{1} + \binom{n}{3}5 + \binom{n}{5}5^2 \ldots\right] + \left[\binom{n}{0} + \binom{n}{2} + \binom{n}{4} + \ldots\right].$$

7. Let $Q = \begin{pmatrix} 1 & 1 \\ 1 & 0 \end{pmatrix}$. Prove:

   (a) $Q^n = \begin{pmatrix} F_{n+1} & F_n \\ F_n & F_{n-1} \end{pmatrix}$.

   (b) $Q^2 = Q + I$; use Q^{2n} to prove the identity in Problem 4.

   (c) Use $Q^{2n+1} = Q^n Q^{n+1}$ to show $F_{2n+1} = F_{n+1}^2 + F_n^2$ and other identities.

8. Show that if $x^2 = x + 1$ then $x^n = F_n x + F_{n-1}$ for $n \geq 2$. Use this result to prove the Binet Formula, namely that $F_n = \frac{\alpha^n - \beta^n}{\alpha - \beta}$.

9. Investigate Lucas, Tribonacci, Tetranacci numbers.

References: The Fibonacci Quarterly; The Golden Section by Garth E. Runion; The Divine Proportion by H.E. Huntley

Project 4: Generalized Pascal Triangles

1. Expand $(1+x+x^2)^2$, $(1+x+x^2)^3$, and $(1+x+x^2)^4$.
2. Organize the coefficients as in the following triangle:

    ```
                    1
                 1  1  1
              1  2  3  2  1
           1  3  6  7  6  3  1
        1  4 10 16 19 16 10  4  1
    ```

3. Investigate the properties of this triangle:

    (a) Produce the next row.

    (b) Give a recursion that produces elements in this triangle.

    (c) What are the row sums? Prove it!

    (d) What is the sum $1^2 + 2^2 + 3^2 + 2^2 + 1^2$? Generalize and prove.

    (e) Is there a hockey stick theorem?

4. Investigate 3-dimensional versions of $(1+x+x^2)^n$ and $(x+y+z)^n$.
5. Investigate $(1+x+x^2+x^3)^n$.
6. Investigate $(1+x+x^2+\ldots+x^k)^n$.
7. Where do the "Fibonacci" numbers get involved in the Pascal triangle?
8. Where do the "Tribonacci" numbers appear in the triangle in Problem 2?

Project 5: A Pascal-like Triangle of Eulerian Numbers

1. Show algebraically and geometrically that $\binom{n}{2} + \binom{n+1}{2} = n^2$.
2. Show that $\binom{n}{3} + 4\binom{n+1}{3} + \binom{n+2}{3} = n^3$.
3. Use Problem 1 and the hockey stick theorem to find a nice formula for $1^2 + 2^2 + 3^2 + \ldots + n^2$. Contrast with the version usually seen in calculus.
4. Use Problem 2 to find a nice formula for $1^3 + 2^3 + 3^3 + \ldots + n^3$.
5. Express n^4 as in Problems 1 and 2.
6. Use Problem 5 to find a nice formula for $1^4 + 2^4 + 3^4 + \ldots + n^4$.
7. Here is an alternate way of accomplishing Problem 6: Find integers $a, b, c,$ and d so that

$$1^4 + 2^4 + 3^4 + \ldots + n^4 = a\binom{n+1}{5} + b\binom{n+2}{5} + c\binom{n+3}{5} + d\binom{n+4}{5}$$

using $n = 1, 2, 3, 4$. This sometimes referred to as "Polynomial Fitting."

8. Express n^5 as in Problems 1 and 2.
9. Express $1^5 + 2^5 + 3^5 + \ldots + n^5$ as a sum of multiples of binomial coefficients.
10. In Problems 1, 2, 5, and 7, n^2, n^3, n^4, and n^5 were expressed as sums of binomial coefficients with integer coefficients. Arrange these expressions in a Pascal-like triangle, concentrating on these integer coefficients. The coefficients are called Eulerian numbers.
11. Investigate the properties of this triangle. Include a discussion of row sums, how entries are found, and a possible recursion. Use $\begin{bmatrix} n \\ k \end{bmatrix}$ for notation.

Project 6: The Leibniz Harmonic Triangle

$$\frac{1}{1}$$
$$\frac{1}{2} \quad \frac{1}{2}$$
$$\frac{1}{3} \quad \frac{1}{6} \quad \frac{1}{3}$$
$$\frac{1}{4} \quad \frac{1}{12} \quad \frac{1}{12} \quad \frac{1}{4}$$
$$\frac{1}{5} \quad \frac{1}{20} \quad \frac{1}{30} \quad \frac{1}{20} \quad \frac{1}{5}$$

1. State the rule used to form each entry, and produce the next two rows.

2. What are the "initial conditions" needed to generate the entries?

3. Use $\begin{bmatrix} n \\ k \end{bmatrix}$ as the notation for entries and state properties analogous to those found in the Pascal Triangle. Include closed formulas for each entry, row sums (alternating also), hockey stick theorem, hexagon property, sum of squares of row entries, etc. Indicate how partial fractions help in verifying the (infinite) hockey stick theorem.

4. See if you can find any papers on this topic. Check *Pi Mu Epsilon Journal, The Mathematical Gazette, Mathematics Magazine, Delta-Undergraduate Mathematics Journal*, etc.

Project 7: The Twelve Days of Christmas (in the Discrete Math Style)

According to the song, *The Twelve Days Of Christmas* the "true love" received the following number of (not so practical) gifts:

Day's Gift	Number received	Total
First	1×12	12
Second	2×11	22
Third	3×10	30
Fourth	4×9	36
Fifth	5×8	40
Sixth	6×7	42
Seventh	7×6	42
Eighth	8×5	40
Ninth	9×4	36
Tenth	10×3	30
Eleventh	11×2	22
Twelfth	12×1	12
		Grand Total = 364

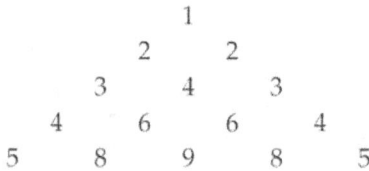

The first five rows of a triangle are given above.

1. Make the next few rows.
2. Where have you seen this triangle before?
3. Experiment with factoring each integer in the 6th row. Repeat with other rows.
4. What are the entries in row 12?
5. Where does the expression $1 \cdot n + 2(n-1) + 3(n-2) + \cdots + n \cdot 1$ show up in the triangle?
6. Explain why $1 \cdot n + 2(n-1) + 3(n-2) + \cdots + n \cdot 1 = \binom{n+2}{3}$.

Appendix B
Background Material

B.1 Mathematical Statements

> **Puzzle 331**
> While walking through a fictional forest, you encounter three trolls guarding a bridge. Each is either a *knight*, who always tells the truth, or a *knave*, who always lies. The trolls will not let you pass until you correctly identify each as either a knight or a knave. Each troll makes a single statement:
>
> Troll 1: If I am a knave, then there are exactly two knights here.
>
> Troll 2: Troll 1 is lying.
>
> Troll 3: Either we are all knaves or at least one of us is a knight.
>
> Which troll is which?

In order to *do* mathematics, we must be able to *talk* and *write* about mathematics. Perhaps your experience with mathematics so far has mostly involved finding answers to problems. As we embark towards more advanced and abstract mathematics, writing will play a more prominent role in the mathematical process.

Communication in mathematics requires more precision than many other subjects, and thus we should take a few pages here to consider the basic building blocks: *mathematical statements*.

B.1.1 Atomic and Molecular Statements

A **statement** is any declarative sentence which is either true or false. A statement is **atomic** if it cannot be divided into smaller statements, otherwise it is called **molecular**.

> **Example B.1.1**
> These are statements (in fact, *atomic* statements):
>
> - Telephone numbers in the USA have 10 digits.
> - The moon is made of cheese.
> - 42 is a perfect square.
> - Every even number greater than 2 can be expressed as the sum of two primes.
> - $3 + 7 = 12$

And these are not statements:

- Would you like some cake?
- The sum of two squares.
- $1 + 3 + 5 + 7 + \cdots + 2n + 1$.
- Go to your room!
- $3 + x = 12$

The reason the sentence "$3 + x = 12$" is not a statement is that it contains a variable. Depending on what x is, the sentence is either true or false, but right now it is neither. One way to make the *sentence* into a *statement* is to specify the value of the variable in some way. This could be done by specifying a specific substitution, for example, "$3 + x = 12$ where $x = 9$," which is a true statement. Or you could *capture* the free variable by *quantifying* over it, as in, "for all values of x, $3 + x = 12$," which is false. We will discuss quantifiers in more detail at the end of this section.

You can build more complicated (molecular) statements out of simpler (atomic or molecular) ones using **logical connectives**. For example, this is a molecular statement:

Telephone numbers in the USA have 10 digits and 42 is a perfect square.

Note that we can break this down into two smaller statements. The two shorter statements are *connected* by an "and." We will consider 5 connectives: "and" (Sam is a man and Chris is a woman), "or" (Sam is a man or Chris is a woman), "if..., then..." (if Sam is a man, then Chris is a woman), "if and only if" (Sam is a man if and only if Chris is a woman), and "not" (Sam is not a man). The first four are called **binary connectives** (because they connect two statements) while "not" is an example of a **unary connective** (since it applies to a single statement).

These molecular statements are of course still statements, so they must be either true or false. The absolutely key observation here is that which **truth value** the molecular statement achieves is completely determined by the type of connective and the truth values of the parts. We do not need to know what the parts actually say, only whether those parts are true or false. So to analyze logical connectives, it is enough to consider **propositional variables** (sometimes called *sentential* variables), usually capital letters in the middle of the alphabet: $P, Q, R, S, \ldots$. We think of these as standing in for (usually atomic) statements, but there are only two *values* the variables can achieve: true or false.[1] We also have symbols for the logical connectives: $\wedge, \vee, \rightarrow, \leftrightarrow, \neg$.

Logical Connectives.

- $P \wedge Q$ is read "P and Q," and called a **conjunction**.
- $P \vee Q$ is read "P or Q," and called a **disjunction**.

---

[1] In computer programing, we should call such variables **Boolean variables**.

- $P \to Q$ is read "if P then Q," and called an **implication** or **conditional**.

- $P \leftrightarrow Q$ is read "P if and only if Q," and called a **biconditional**.

- $\neg P$ is read "not P," and called a **negation**.

The **truth value** of a statement is determined by the truth value(s) of its part(s), depending on the connectives:

Truth Conditions for Connectives.

- $P \wedge Q$ is true when both P and Q are true

- $P \vee Q$ is true when P or Q or both are true.

- $P \to Q$ is true when P is false or Q is true or both.

- $P \leftrightarrow Q$ is true when P and Q are both true, or both false.

- $\neg P$ is true when P is false.

Note that for us, *or* is the **inclusive or** (and not the sometimes used *exclusive or*) meaning that $P \vee Q$ is in fact true when both P and Q are true. As for the other connectives, "and" behaves as you would expect, as does negation. The biconditional (if and only if) might seem a little strange, but you should think of this as saying the two parts of the statements are *equivalent* in that they have the same truth value. This leaves only the conditional $P \to Q$ which has a slightly different meaning in mathematics than it does in ordinary usage. However, implications are so common and useful in mathematics, that we must develop fluency with their use, and as such, they deserve their own subsection.

B.1.2 Implications

Implications.

An **implication** or **conditional** is a molecular statement of the form

$$P \to Q$$

where P and Q are statements. We say that

- P is the **hypothesis** (or **antecedent**).

- Q is the **conclusion** (or **consequent**).

An implication is *true* provided P is false or Q is true (or both), and *false* otherwise. In particular, the only way for $P \to Q$ to be false is for P to be true *and* Q to be false.

Easily the most common type of statement in mathematics is the implication. Even statements that do not at first look like they have this form conceal an implication at their heart. Consider the *Pythagorean Theorem*. Many a college freshman would quote this theorem as "$a^2 + b^2 = c^2$." This is absolutely not correct. For one thing, that is not a statement since it has three variables in it. Perhaps they imply that this should be true for any values of the variables? So $1^2 + 5^2 = 2^2$??? How can we fix this? Well, the equation is true as long as a and b are the legs or a right triangle and c is the hypotenuse. In other words:

If a and b are the legs of a right triangle with hypotenuse c, then $a^2 + b^2 = c^2$.

This is a reasonable way to think about implications: our claim is that the conclusion ("then" part) is true, but on the assumption that the hypothesis ("if" part) is true. We make no claim about the conclusion in situations when the hypothesis is false.[2]

Still, it is important to remember that an implication is a statement, and therefore is either true or false. The truth value of the implication is determined by the truth values of its two parts. To agree with the usage above, we say that an implication is true either when the hypothesis is false, or when the conclusion is true. This leaves only one way for an implication to be false: when the hypothesis is true and the conclusion is false.

Example B.1.2
Consider the statement:

If Bob gets a 90 on the final, then Bob will pass the class.

This is definitely an implication: P is the statement "Bob gets a 90 on the final," and Q is the statement "Bob will pass the class."

Suppose I made that statement to Bob. In what circumstances would it be fair to call me a liar? What if Bob really did get a 90 on the final, and he did pass the class? Then I have not lied; my statement is true. However, if Bob did get a 90 on the final and did not pass the class, then I lied, making the statement false. The tricky case is this: what if Bob did not get a 90 on the final? Maybe he passes the class, maybe he doesn't. Did I lie in either case? I think not. In these last two cases, P was false, and the statement $P \to Q$ was true. In the first case, Q was true, and so was $P \to Q$. So $P \to Q$ is true when either P is false or Q is true.

Just to be clear, although we sometimes read $P \to Q$ as "*P implies Q*", we are not insisting that there is some causal relationship between the statements P and Q. In particular, if you claim that $P \to Q$ is *false*, you are not saying that P does not imply Q, but rather that P is true and Q is false.

Example B.1.3
Decide which of the following statements are true and which are false. Briefly explain.

---

[2]However, note that in the case of the Pythagorean Theorem, it is also the case that *if $a^2 + b^2 = c^2$, then a and b are the legs of a right triangle with hypotenuse c.* So we could have also expressed this theorem as a biconditional: "a and b are the legs of a right triangle with hypotenuse c *if and only if* $a^2 + b^2 = c^2$."

1. If $1 = 1$, then most horses have 4 legs.

2. If $0 = 1$, then $1 = 1$.

3. If 8 is a prime number, then the 7624th digit of π is an 8.

4. If the 7624th digit of π is an 8, then $2 + 2 = 4$.

Solution. All four of the statements are true. Remember, the only way for an implication to be false is for the *if* part to be true and the *then* part to be false.

1. Here both the hypothesis and the conclusion are true, so the implication is true. It does not matter that there is no meaningful connection between the true mathematical fact and the fact about horses.

2. Here the hypothesis is false and the conclusion is true, so the implication is true.

3. I have no idea what the 7624th digit of π is, but this does not matter. Since the hypothesis is false, the implication is automatically true.

4. Similarly here, regardless of the truth value of the hypothesis, the conclusion is true, making the implication true.

It is important to understand the conditions under which an implication is true not only to decide whether a mathematical statement is true, but in order to *prove* that it is. Proofs might seem scary (especially if you have had a bad high school geometry experience) but all we are really doing is explaining (very carefully) why a statement is true. If you understand the truth conditions for an implication, you already have the outline for a proof.

Direct Proofs of Implications.

To prove an implication $P \to Q$, it is enough to assume P, and from it, deduce Q.

Perhaps a better way to say this is that to prove a statement of the form $P \to Q$ directly, you must explain why Q is true, but you *get to* assume P is true first. After all, you only care about whether Q is true in the case that P is as well.

There are other techniques to prove statements (implications and others) that we will encounter throughout our studies, and new proof techniques are discovered all the time. Direct proof is the easiest and most elegant style of proof and has the advantage that such a proof often does a great job of explaining *why* the statement is true.

Example B.1.4
Prove: If two numbers a and b are even, then their sum $a + b$ is even.
Solution.

Proof. Suppose the numbers a and b are even. This means that $a = 2k$ and $b = 2j$ for some integers k and j. The sum is then $a+b = 2k+2j = 2(k+j)$. Since $k + j$ is an integer, this means that $a + b$ is even. ∎

Notice that since we get to assume the hypothesis of the implication, we immediately have a place to start. The proof proceeds essentially by repeatedly asking and answering, "what does that mean?" Eventually, we conclude that it means the conclusion.

This sort of argument shows up outside of math as well. If you ever found yourself starting an argument with "hypothetically, let's assume ...," then you have attempted a direct proof of your desired conclusion.

An implication is a way of expressing a relationship between two statements. It is often interesting to ask whether there are other relationships between the statements. Here we introduce some common language to address this question.

Converse and Contrapositive.

- The **converse** of an implication $P \to Q$ is the implication $Q \to P$. The converse is NOT logically equivalent to the original implication. That is, whether the converse of an implication is true is independent of the truth of the implication.

- The **contrapositive** of an implication $P \to Q$ is the statement $\neg Q \to \neg P$. An implication and its contrapositive are logically equivalent (they are either both true or both false).

Mathematics is overflowing with examples of true implications which have a false converse. If a number greater than 2 is prime, then that number is odd. However, just because a number is odd does not mean it is prime. If a shape is a square, then it is a rectangle. But it is false that if a shape is a rectangle, then it is a square.

However, sometimes the converse of a true statement is also true. For example, the Pythagorean theorem has a true converse: if $a^2 + b^2 = c^2$, then the triangle with sides a, b, and c is a *right* triangle. Whenever you encounter an implication in mathematics, it is always reasonable to ask whether the converse is true.

The contrapositive, on the other hand, always has the same truth value as its original implication. This can be very helpful in deciding whether an implication is true: often it is easier to analyze the contrapositive.

Example B.1.5

True or false: If you draw any nine playing cards from a regular deck, then you will have at least three cards all of the same suit. Is the converse true?

Solution. True. The original implication is a little hard to analyze because there are so many different combinations of nine cards. But consider the contrapositive: If you *don't* have at least three cards all of the same suit, then you don't have nine cards. It is easy to see why this is true: you can at most have two cards of each of the four suits, for a total of eight cards (or fewer).

The converse: If you have at least three cards all of the same suit, then you have nine cards. This is false. You could have three spades and nothing else. Note that to demonstrate that the converse (an implication) is false, we provided an example where the hypothesis is true (you do have three cards of the same suit), but where the conclusion is false (you do not have nine cards).

Understanding converses and contrapositives can help understand implications and their truth values:

Example B.1.6
Suppose I tell Sue that if she gets a 93% on her final, then she will get an A in the class. Assuming that what I said is true, what can you conclude in the following cases:

1. Sue gets a 93% on her final.

2. Sue gets an A in the class.

3. Sue does not get a 93% on her final.

4. Sue does not get an A in the class.

Solution. Note first that whenever $P \to Q$ and P are both true statements, Q must be true as well. For this problem, take P to mean "Sue gets a 93% on her final" and Q to mean "Sue will get an A in the class."

1. We have $P \to Q$ and P, so Q follows. Sue gets an A.

2. You cannot conclude anything. Sue could have gotten the A because she did extra credit for example. Notice that we do not know that if Sue gets an A, then she gets a 93% on her final. That is the converse of the original implication, so it might or might not be true.

3. The contrapositive of the converse of $P \to Q$ is $\neg P \to \neg Q$, which states that if Sue does not get a 93% on the final, then she will not get an A in the class. But this does not follow from the original implication. Again, we can conclude nothing. Sue could have done extra credit.

4. What would happen if Sue does not get an A but *did* get a 93% on the final? Then P would be true and Q would be false. This makes the implication $P \to Q$ false! It must be that Sue did not get a 93% on the final. Notice now we have the implication $\neg Q \to \neg P$ which is the contrapositive of $P \to Q$. Since $P \to Q$ is assumed to be true, we know $\neg Q \to \neg P$ is true as well.

As we said above, an implication is not logically equivalent to its converse, but it is possible that both the implication and its converse are true. In this case, when both $P \to Q$ and $Q \to P$ are true, we say that P and Q are equivalent and write $P \leftrightarrow Q$. This is the biconditional we mentioned earlier.

> **If and only if.**
>
> $P \leftrightarrow Q$ is logically equivalent to $(P \to Q) \land (Q \to P)$.
>
> Example: Given an integer n, it is true that n is even if and only if n^2 is even. That is, if n is even, then n^2 is even, as well as the converse: if n^2 is even, then n is even.

You can think of "if and only if" statements as having two parts: an implication and its converse. We might say one is the "if" part, and the other is the "only if" part. We also sometimes say that "if and only if" statements have two directions: a forward direction ($P \to Q$) and a backwards direction ($P \leftarrow Q$, which is really just sloppy notation for $Q \to P$).

Let's think a little about which part is which. Is $P \to Q$ the "if" part or the "only if" part? Consider an example.

> **Example B.1.7**
>
> Suppose it is true that I sing if and only if I'm in the shower. We know this means both that if I sing, then I'm in the shower, and also the converse, that if I'm in the shower, then I sing. Let P be the statement, "I sing," and Q be, "I'm in the shower." So $P \to Q$ is the statement "if I sing, then I'm in the shower." Which part of the if and only if statement is this?
>
> What we are really asking for is the meaning of "I sing *if* I'm in the shower" and "I sing *only if* I'm in the shower." When is the first one (the "if" part) *false*? When I am in the shower but not singing. That is the same condition on being false as the statement "if I'm in the shower, then I sing." So the "if" part is $Q \to P$. On the other hand, to say, "I sing only if I'm in the shower" is equivalent to saying "if I sing, then I'm in the shower," so the "only if" part is $P \to Q$.

It is not terribly important to know which part is the "if" or "only if" part, but this does illustrate something very, very important: *there are many ways to state an implication!*

> **Example B.1.8**
>
> Rephrase the implication, "if I dream, then I am asleep" in as many different ways as possible. Then do the same for the converse.
>
> **Solution.** The following are all equivalent to the original implication:
>
> 1. I am asleep if I dream.
>
> 2. I dream only if I am asleep.
>
> 3. In order to dream, I must be asleep.

4. To dream, it is necessary that I am asleep.

5. To be asleep, it is sufficient to dream.

6. I am not dreaming unless I am asleep.

The following are equivalent to the converse (if I am asleep, then I dream):

1. I dream if I am asleep.

2. I am asleep only if I dream.

3. It is necessary that I dream in order to be asleep.

4. It is sufficient that I be asleep in order to dream.

5. If I don't dream, then I'm not asleep.

Hopefully you agree with the above example. We include the "necessary and sufficient" versions because those are common when discussing mathematics. In fact, let's agree once and for all what they mean.

Necessary and Sufficient.

- "P is necessary for Q" means $Q \to P$.

- "P is sufficient for Q" means $P \to Q$.

- If P is necessary and sufficient for Q, then $P \leftrightarrow Q$.

To be honest, I have trouble with these if I'm not very careful. I find it helps to keep a standard example for reference.

Example B.1.9

Recall from calculus, if a function is differentiable at a point c, then it is continuous at c, but that the converse of this statement is not true (for example, $f(x) = |x|$ at the point 0). Restate this fact using "necessary and sufficient" language.

Solution. It is true that in order for a function to be differentiable at a point c, it is necessary for the function to be continuous at c. However, it is not necessary that a function be differentiable at c for it to be continuous at c.

It is true that to be continuous at a point c, it is sufficient that the function be differentiable at c. However, it is not the case that being continuous at c is sufficient for a function to be differentiable at c.

Thinking about the necessity and sufficiency of conditions can also help when writing proofs and justifying conclusions. If you want to establish some mathematical fact, it is helpful to think what other facts would *be enough* (be

B.1.3 Predicates and Quantifiers

Puzzle 332

Consider the statements below. Decide whether any are equivalent to each other, or whether any imply any others.

1. You can fool some people all of the time.
2. You can fool everyone some of the time.
3. You can always fool some people.
4. Sometimes you can fool everyone.

It would be nice to use variables in our mathematical sentences. For example, suppose we wanted to claim that if n is prime, then $n + 7$ is not prime. This looks like an implication. I would like to write something like

$$P(n) \rightarrow \neg P(n + 7)$$

where $P(n)$ means "n is prime." But this is not quite right. For one thing, because this sentence has a **free variable** (that is, a variable that we have not specified anything about), it is not a statement. A sentence that contains variables is called a **predicate**.

Now, if we plug in a specific value for n, we do get a statement. In fact, it turns out that no matter what value we plug in for n, we get a true implication in this case. What we really want to say is that *for all* values of n, if n is prime, then $n + 7$ is not. We need to *quantify* the variable.

Although there are many types of *quantifiers* in English (e.g., many, few, most, etc.) in mathematics we, for the most part, stick to two: existential and universal.

Universal and Existential Quantifiers.

The existential quantifier is $\exists$ and is read "there exists" or "there is." For example,

$$\exists x (x < 0)$$

asserts that there is a number less than 0.

The universal quantifier is $\forall$ and is read "for all" or "every." For example,

$$\forall x (x \geq 0)$$

asserts that every number is greater than or equal to 0.

As with all mathematical statements, we would like to decide whether quantified statements are true or false. Consider the statement

$$\forall x \exists y (y < x).$$

You would read this, "for every x there is some y such that y is less than x." Is this true? The answer depends on what our *domain of discourse* is: when we say "for all" x, do we mean all positive integers or all real numbers or all elements of some other set? Usually this information is implied. In discrete mathematics, we almost always quantify over the *natural numbers*, 0, 1, 2, ..., so let's take that for our domain of discourse here.

For the statement to be true, we need it to be the case that no matter what natural number we select, there is always some natural number that is strictly smaller. Perhaps we could let y be $x - 1$? But here is the problem: what if $x = 0$? Then $y = -1$ and that is *not a number!* (in our domain of discourse). Thus we see that the statement is false because there is a number which is less than or equal to all other numbers. In symbols,

$$\exists x \forall y (y \geq x).$$

To show that the original statement is false, we proved that the *negation* was true. Notice how the negation and original statement compare. This is typical.

> **Quantifiers and Negation.**
>
> $\neg \forall x P(x)$ is equivalent to $\exists x \neg P(x)$.
>
> $\neg \exists x P(x)$ is equivalent to $\forall x \neg P(x)$.

Essentially, we can pass the negation symbol over a quantifier, but that causes the quantifier to switch type. This should not be surprising: if not everything has a property, then something doesn't have that property. And if there is not something with a property, then everything doesn't have that property.

Implicit Quantifiers.

It is always a good idea to be precise in mathematics. Sometimes though, we can relax a little bit, as long as we all agree on a convention. An example of such a convention is to assume that sentences containing predicates with free variables are intended as statements, where the variables are universally quantified.

For example, do you believe that if a shape is a square, then it is a rectangle? But how can that be true if it is not a statement? To be a little more precise, we have two predicates: $S(x)$ standing for "x is a square" and $R(x)$ standing for "x is a rectangle". The *sentence* we are looking at is,

$$S(x) \rightarrow R(x).$$

This is neither true nor false, as it is not a statement. But come on! We all know that we meant to consider the statement,

$$\forall x (S(x) \rightarrow R(x)),$$

and this is what our convention tells us to consider.

Similarly, we will often be a bit sloppy about the distinction between a predicate and a statement. For example, we might write, *let $P(n)$ be the* statement, "*n is prime,*" which is technically incorrect. It is implicit that we mean that we are defining $P(n)$ to be a predicate, which for each n becomes the statement, n is prime.

B.1.4 Exercises

1. For each sentence below, decide whether it is an atomic statement, a molecular statement, or not a statement at all.

    (a) Customers must wear shoes.

    (b) The customers wore shoes.

    (c) The customers wore shoes and they wore socks.

    [Solution]

2. Classify each of the sentences below as an atomic statement, a molecular statement, or not a statement at all. If the statement is molecular, say what kind it is (conjuction, disjunction, conditional, biconditional, negation).

    (a) The sum of the first 100 odd positive integers.

    (b) Everybody needs somebody sometime.

    (c) The Broncos will win the Super Bowl or I'll eat my hat.

    (d) We can have donuts for dinner, but only if it rains.

    (e) Every natural number greater than 1 is either prime or composite.

    (f) This sentence is false.

3. Suppose P and Q are the statements: P: Jack passed math. Q: Jill passed math.

    (a) Translate "Jack and Jill both passed math" into symbols.

    (b) Translate "If Jack passed math, then Jill did not" into symbols.

    (c) Translate "$P \vee Q$" into English.

    (d) Translate "$\neg(P \wedge Q) \rightarrow Q$" into English.

    (e) Suppose you know that if Jack passed math, then so did Jill. What can you conclude if you know that:

        i. Jill passed math?
        ii. Jill did not pass math?

    [Solution]

4. Determine whether each molecular statement below is true or false, or whether it is impossible to determine. Assume you do not know what my favorite number is (but you do know that 13 is prime).

    (a) If 13 is prime, then 13 is my favorite number.

    (b) If 13 is my favorite number, then 13 is prime.

    (c) If 13 is not prime, then 13 is my favorite number.

(d) 13 is my favorite number or 13 is prime.

(e) 13 is my favorite number and 13 is prime.

(f) 7 is my favorite number and 13 is not prime.

(g) 13 is my favorite number or 13 is not my favorite number.

[Solution]

5. In my safe is a sheet of paper with two shapes drawn on it in colored crayon. One is a square, and the other is a triangle. Each shape is drawn in a single color. Suppose you believe me when I tell you that *if the square is blue, then the triangle is green*. What do you therefore know about the truth value of the following statements?

    (a) The square and the triangle are both blue.

    (b) The square and the triangle are both green.

    (c) If the triangle is not green, then the square is not blue.

    (d) If the triangle is green, then the square is blue.

    (e) The square is not blue or the triangle is green.

[Solution]

6. Again, suppose the statement "if the square is blue, then the triangle is green" is true. This time however, assume the converse is false. Classify each statement below as true or false (if possible).

    (a) The square is blue if and only if the triangle is green.

    (b) The square is blue if and only if the triangle is not green.

    (c) The square is blue.

    (d) The triangle is green.

[Solution]

7. Consider the statement, "If you will give me a cow, then I will give you magic beans." Decide whether each statement below is the converse, the contrapositive, or neither.

    (a) If you will give me a cow, then I will not give you magic beans.

    (b) If I will not give you magic beans, then you will not give me a cow.

    (c) If I will give you magic beans, then you will give me a cow.

    (d) If you will not give me a cow, then I will not give you magic beans.

    (e) You will give me a cow and I will not give you magic beans.

    (f) If I will give you magic beans, then you will not give me a cow.

[Solution]

8. Consider the statement "If Oscar eats Chinese food, then he drinks milk."

   (a) Write the converse of the statement.

   (b) Write the contrapositive of the statement.

   (c) Is it possible for the contrapositive to be false? If it was, what would that tell you?

   (d) Suppose the original statement is true, and that Oscar drinks milk. Can you conclude anything (about his eating Chinese food)? Explain.

   (e) Suppose the original statement is true, and that Oscar does not drink milk. Can you conclude anything (about his eating Chinese food)? Explain.

9. You have discovered an old paper on graph theory that discusses the *viscosity* of a graph (which for all you know, is something completely made up by the author). A theorem in the paper claims that "if a graph satisfies *condition (V)*, then the graph is *viscous*." Which of the following are equivalent ways of stating this claim? Which are equivalent to the *converse* of the claim?

   (a) A graph is viscous only if it satisfies condition (V).

   (b) A graph is viscous if it satisfies condition (V).

   (c) For a graph to be viscous, it is necessary that it satisfies condition (V).

   (d) For a graph to be viscous, it is sufficient for it to satisfy condition (V).

   (e) Satisfying condition (V) is a sufficient condition for a graph to be viscous.

   (f) Satisfying condition (V) is a necessary condition for a graph to be viscous.

   (g) Every viscous graph satisfies condition (V).

   (h) Only viscous graphs satisfy condition (V).

   [Solution]

10. Write each of the following statements in the form, "if..., then...." Careful, some of the statements might be false (which is alright for the purposes of this question).

    (a) To lose weight, you must exercise.

    (b) To lose weight, all you need to do is exercise.

    (c) Every American is patriotic.

    (d) You are patriotic only if you are American.

    (e) The set of rational numbers is a subset of the real numbers.

    (f) A number is prime if it is not even.

    (g) Either the Broncos will win the Super Bowl, or they won't play in the Super Bowl.

[Solution]

11. Which of the following statements are equivalent to the implication, "if you win the lottery, then you will be rich," and which are equivalent to the converse of the implication?

    (a) Either you win the lottery or else you are not rich.

    (b) Either you don't win the lottery or else you are rich.

    (c) You will win the lottery and be rich.

    (d) You will be rich if you win the lottery.

    (e) You will win the lottery if you are rich.

    (f) It is necessary for you to win the lottery to be rich.

    (g) It is sufficient to win the lottery to be rich.

    (h) You will be rich only if you win the lottery.

    (i) Unless you win the lottery, you won't be rich.

    (j) If you are rich, you must have won the lottery.

    (k) If you are not rich, then you did not win the lottery.

    (l) You will win the lottery if and only if you are rich.

12. Let $P(x)$ be the predicate, "$3x + 1$ is even."

    (a) Is $P(5)$ true or false?

    (b) What, if anything, can you conclude about $\exists x P(x)$ from the truth value of $P(5)$?

    (c) What, if anything, can you conclude about $\forall x P(x)$ from the truth value of $P(5)$?

[Solution]

13. Let $P(x)$ be the predicate, "$4x + 1$ is even."

    (a) Is $P(5)$ true or false?

    (b) What, if anything, can you conclude about $\exists x P(x)$ from the truth value of $P(5)$?

    (c) What, if anything, can you conclude about $\forall x P(x)$ from the truth value of $P(5)$?

14. For a given predicate $P(x)$, you might believe that the statements $\forall x P(x)$ or $\exists x P(x)$ are either true or false. How would you decide if you were correct in each case? You have four choices: you could give an example of an element n in the domain for which $P(n)$ is true or for which $P(n)$ if false, or you could argue that no matter what n is, $P(n)$ is true or is false.

    (a) What would you need to do to prove $\forall x P(x)$ is true?

    (b) What would you need to do to prove $\forall x P(x)$ is false?

(c) What would you need to do to prove $\exists x P(x)$ is true?

(d) What would you need to do to prove $\exists x P(x)$ is false?

[Solution]

15. Suppose $P(x, y)$ is some binary predicate defined on a very small domain of discourse: just the integers 1, 2, 3, and 4. For each of the 16 pairs of these numbers, $P(x, y)$ is either true or false, according to the following table (x values are rows, y values are columns).

	1	2	3	4
1	T	F	F	F
2	F	T	T	F
3	T	T	T	T
4	F	F	F	F

For example, $P(1, 3)$ is false, as indicated by the F in the first row, third column.

Use the table to decide whether the following statements are true or false.

(a) $\forall x \exists y P(x, y)$.

(b) $\forall y \exists x P(x, y)$.

(c) $\exists x \forall y P(x, y)$.

(d) $\exists y \forall x P(x, y)$.

[Solution]

16. Translate into symbols. Use $E(x)$ for "x is even" and $O(x)$ for "x is odd."

(a) No number is both even and odd.

(b) One more than any even number is an odd number.

(c) There is prime number that is even.

(d) Between any two numbers there is a third number.

(e) There is no number between a number and one more than that number.

[Solution]

17. Translate into English:

(a) $\forall x (E(x) \to E(x+2))$.

(b) $\forall x \exists y (\sin(x) = y)$.

(c) $\forall y \exists x (\sin(x) = y)$.

(d) $\forall x \forall y (x^3 = y^3 \to x = y)$.

[Solution]

18. Suppose $P(x)$ is some predicate for which the statement $\forall x P(x)$ is true. Is it also the case that $\exists x P(x)$ is true? In other words, is the statement $\forall x P(x) \to \exists x P(x)$ always true? Is the converse always true? Assume the domain of discourse is non-empty.

    Hint. Try an example. What if $P(x)$ was the predicate, "x is prime"? What if it was "if x is divisible by 4, then it is even"? Of course examples are not enough to prove something in general, but that is entirely the point of this question.

19. For each of the statements below, give a domain of discourse for which the statement is true, and a domain for which the statement is false.

    (a) $\forall x \exists y (y^2 = x)$.

    (b) $\forall x \forall y (x < y \to \exists z (x < z < y))$.

    (c) $\exists x \forall y \forall z (y < z \to y \le x \le z)$.

    Hint. First figure out what each statement is saying. For part (c), you don't need to assume the domain is an infinite set.

20. Consider the statement, "For all natural numbers n, if n is prime, then n is solitary." You do not need to know what *solitary* means for this problem, just that it is a property that some numbers have and others do not.

    (a) Write the converse and the contrapositive of the statement, saying which is which. Note: the original statement claims that an implication is true for all n, and it is that implication that we are taking the converse and contrapositive of.

    (b) Write the negation of the original statement. What would you need to show to prove that the statement is false?

    (c) Even though you don't know whether 10 is solitary (in fact, nobody knows this), is the statement "if 10 is prime, then 10 is solitary" true or false? Explain.

    (d) It turns out that 8 is solitary. Does this tell you anything about the truth or falsity of the original statement, its converse or its contrapositive? Explain.

    (e) Assuming that the original statement is true, what can you say about the relationship between the *set P* of prime numbers and the *set S* of solitary numbers. Explain.

B.2 Sets

The most fundamental objects we will use in our studies (and really in all of math) are *sets*. Much of what follows might be review, but it is very important that you are fluent in the language of set theory. Most of the notation we use below is standard, although some might be a little different than what you have seen before.

For us, a **set** will simply be an unordered collection of objects. Two examples: we could consider the set of all actors who have played *The Doctor* on *Doctor Who*, or the set of natural numbers between 1 and 10 inclusive. In the first case, Tom Baker is a element (or member) of the set, while Idris Elba, among many others, is not an element of the set. Also, the two examples are of different sets. Two sets are equal exactly if they contain the exact same elements. For example, the set containing all of the vowels in the declaration of independence is precisely the same set as the set of vowels in the word "questionably" (namely, all of them); we do not care about order or repetitions, just whether the element is in the set or not.

B.2.1 Notation

We need some notation to make talking about sets easier. Consider,

$$A = \{1, 2, 3\}.$$

This is read, "A is the set containing the elements 1, 2 and 3." We use curly braces "{, }" to enclose elements of a set. Some more notation:

$$a \in \{a, b, c\}.$$

The symbol "$\in$" is read "is in" or "is an element of." Thus the above means that a is an element of the set containing the letters a, b, and c. Note that this is a true statement. It would also be true to say that d is not in that set:

$$d \notin \{a, b, c\}.$$

Be warned: we write "$x \in A$" when we wish to express that one of the elements of the set A is x. For example, consider the set,

$$A = \{1, b, \{x, y, z\}, \emptyset\}.$$

This is a strange set, to be sure. It contains four elements: the number 1, the letter b, the set $\{x, y, z\}$, and the empty set $\emptyset = \{\}$, the set containing no elements. Is x in A? The answer is no. None of the four elements in A are the letter x, so we must conclude that $x \notin A$. Similarly, consider the set $B = \{1, b\}$. Even though the elements of B are elements of A, we cannot say that the *set B* is one of the elements of A. Therefore $B \notin A$. (Soon we will see that B is a *subset* of A, but this is different from being an *element* of A.)

We have described the sets above by listing their elements. Sometimes this is hard to do, especially when there are a lot of elements in the set (perhaps infinitely many). For instance, if we want A to be the set of all even natural numbers, would could write,

$$A = \{0, 2, 4, 6, \ldots\},$$

but this is a little imprecise. A better way would be

$$A = \{x \in \mathbb{N} : \exists n \in \mathbb{N}(x = 2n)\}.$$

Let's look at this carefully. First, there are some new symbols to digest: "$\mathbb{N}$" is the symbol usually used to denote that **natural numbers**, which we will take to be the set $\{0, 1, 2, 3, \ldots\}$. Next, the colon, ":", is read *such that*; it separates the elements that are in the set from the condition that the elements in the set must satisfy. So putting this all together, we would read the set as, "the set of all x in the natural numbers, such that there exists some n in the natural numbers for which x is twice n." In other words, the set of all natural numbers, that are even. Here is another way to write the same set.

$$A = \{x \in \mathbb{N} : x \text{ is even}\}.$$

Note: Sometimes mathematicians use | or ∋ for the "such that" symbol instead of the colon. Also, there is a fairly even split between mathematicians about whether 0 is an element of the natural numbers, so be careful there.

This notation is usually called **set builder notation**. It tells use how to *build* a set by telling us pricisely the condition elements must meet to gain access (the condition is the logical statement after the ":" symbol). Reading and comprehending sets written in this way takes practice. Here are some more examples:

Example B.2.1

Describe each of the following sets both in words and by listing out enough elements to see the pattern.

1. $\{x : x + 3 \in \mathbb{N}\}$.

2. $\{x \in \mathbb{N} : x + 3 \in \mathbb{N}\}$.

3. $\{x : x \in \mathbb{N} \vee -x \in \mathbb{N}\}$.

4. $\{x : x \in \mathbb{N} \wedge -x \in \mathbb{N}\}$.

Solution.

1. This is the set of all numbers which are 3 less than a natural number (i.e., that if you add 3 to them, you get a natural number). The set could also be written as $\{-3, -2, -1, 0, 1, 2, \ldots\}$ (note that 0 is a natural number, so -3 is in this set because $-3 + 3 = 0$).

2. This is the set of all natural numbers which are 3 less than a natural number. So here we just have $\{0, 1, 2, 3 \ldots\}$.

3. This is the set of all integers (positive and negative whole numbers, written $\mathbb{Z}$). In other words, $\{\ldots, -2, -1, 0, 1, 2, \ldots\}$.

4. Here we want all numbers x such that x and $-x$ are natural numbers. There is only one: 0. So we have the set $\{0\}$.

There is also a subtle variation on set builder notation. While the condition is generally given after the "such that", sometimes it is hidden in the first part. Here is an example.

> **Example B.2.2**
> List a few elements in the sets below and describe them in words. The set $\mathbb{Z}$ is the set of **integers**; positive and negative whole numbers.
>
> 1. $A = \{x \in \mathbb{Z} : x^2 \in \mathbb{N}\}$
> 2. $B = \{x^2 : x \in \mathbb{N}\}$
>
> **Solution.**
>
> 1. The set of integers that pass the condition that their square is a natural number. Well, every integer, when you square it, gives you a non-negative integer, so a natural number. Thus $A = \mathbb{Z} = \{\ldots, -2, -1, 0, 1, 2, 3, \ldots\}$.
>
> 2. Here we are looking for the set of all x^2s where x is a natural number. So this set is simply the set of perfect squares. $B = \{0, 1, 4, 9, 16, \ldots\}$.
>
>   Another way we could have written this set, using more strict set builder notation, would be as $B = \{x \in \mathbb{N} : x = n^2 \text{ for some } n \in \mathbb{N}\}$.

We already have a lot of notation, and there is more yet. Below is a handy chart of symbols. Some of these will be discussed in greater detail as we move forward.

Special sets.

$\emptyset$	The **empty set** is the set which contains no elements.
$\mathcal{U}$	The **universe set** is the set of all elements.
$\mathbb{N}$	The set of natural numbers. That is, $\mathbb{N} = \{0, 1, 2, 3 \ldots\}$.
$\mathbb{Z}$	The set of integers. That is, $\mathbb{Z} = \{\ldots, -2, -1, 0, 1, 2, 3, \ldots\}$.
$\mathbb{Q}$	The set of rational numbers.
$\mathbb{R}$	The set of real numbers.
$\mathcal{P}(A)$	The **power set** of any set A is the set of all subsets of A.

Set Theory Notation.

$\{,\}$	We use these **braces** to enclose the elements of a set. So $\{1, 2, 3\}$ is the set containing 1, 2, and 3.
$:$	$\{x : x > 2\}$ is the set of all x **such that** x is greater than 2.
$\in$	$2 \in \{1, 2, 3\}$ asserts that 2 is **an element of** the set $\{1, 2, 3\}$.
$\notin$	$4 \notin \{1, 2, 3\}$ because 4 **is not an element of** the set $\{1, 2, 3\}$.

⊆	$A \subseteq B$ asserts that A **is a subset of** B: every element of A is also an element of B.
⊂	$A \subset B$ asserts that A **is a proper subset of** B: every element of A is also an element of B, but $A \neq B$.
∩	$A \cap B$ is the **intersection of** A **and** B: the set containing all elements which are elements of both A and B.
∪	$A \cup B$ is the **union of** A **and** B: is the set containing all elements which are elements of A or B or both.
×	$A \times B$ is the **Cartesian product of** A **and** B: the set of all ordered pairs (a, b) with $a \in A$ and $b \in B$.
\	$A \setminus B$ is A **set-minus** B: the set containing all elements of A which are not elements of B.
$\bar{A}$	The **complement of** A is the set of everything which is not an element of A.
$|A|$	The **cardinality (or size) of** A is the number of elements in A.

Puzzle 333

1. Find the cardinality of each set below.
   (a) $A = \{3, 4, \ldots, 15\}$.
   (b) $B = \{n \in \mathbb{N} : 2 < n \leq 200\}$.
   (c) $C = \{n \leq 100 : n \in \mathbb{N} \wedge \exists m \in \mathbb{N}(n = 2m + 1)\}$.
2. Find two sets A and B for which $|A| = 5$, $|B| = 6$, and $|A \cup B| = 9$. What is $|A \cap B|$?
3. Find sets A and B with $|A| = |B|$ such that $|A \cup B| = 7$ and $|A \cap B| = 3$. What is $|A|$?
4. Let $A = \{1, 2, \ldots, 10\}$. Define $\mathcal{B}_2 = \{B \subseteq A : |B| = 2\}$. Find $|\mathcal{B}_2|$.
5. For any sets A and B, define $AB = \{ab : a \in A \wedge b \in B\}$. If $A = \{1, 2\}$ and $B = \{2, 3, 4\}$, what is $|AB|$? What is $|A \times B|$?

B.2.2 Relationships Between Sets

We have already said what it means for two sets to be equal: they have exactly the same elements. Thus, for example,

$$\{1, 2, 3\} = \{2, 1, 3\}.$$

(Remember, the order the elements are written down in does not matter.) Also,
$$\{1,2,3\} = \{1, 1+1, 1+1+1\} = \{I, II, III\} = \{1, 2, 3, 1+2\}$$
since these are all ways to write the set containing the first three positive integers (how we write them doesn't matter, just what they are).

What about the sets $A = \{1, 2, 3\}$ and $B = \{1, 2, 3, 4\}$? Clearly $A \neq B$, but notice that every element of A is also an element of B. Because of this we say that A is a *subset* of B, or in symbols $A \subset B$ or $A \subseteq B$. Both symbols are read "is a subset of." The difference is that sometimes we want to say that A is either equal to or is a subset of B, in which case we use $\subseteq$. This is analogous to the difference between $<$ and $\leq$.

Example B.2.3
Let $A = \{1, 2, 3, 4, 5, 6\}$, $B = \{2, 4, 6\}$, $C = \{1, 2, 3\}$ and $D = \{7, 8, 9\}$. Determine which of the following are true, false, or meaningless.

1. $A \subset B$.
2. $B \subset A$.
3. $B \in C$.
4. $\emptyset \in A$.
5. $\emptyset \subset A$.
6. $A < D$.
7. $3 \in C$.
8. $3 \subset C$.
9. $\{3\} \subset C$.

Solution.

1. False. For example, $1 \in A$ but $1 \notin B$.
2. True. Every element in B is an element in A.
3. False. The elements in C are 1, 2, and 3. The *set* B is not equal to 1, 2, or 3.
4. False. A has exactly 6 elements, and none of them are the empty set.
5. True. Everything in the empty set (nothing) is also an element of A. Notice that the empty set is a subset of every set.
6. Meaningless. A set cannot be less than another set.
7. True. 3 is one of the elements of the set C.
8. Meaningless. 3 is not a set, so it cannot be a subset of another set.
9. True. 3 is the only element of the set $\{3\}$, and is an element of C, so every element in $\{3\}$ is an element of C.

In the example above, B is a subset of A. You might wonder what other sets are subsets of A. If you collect all these subsets of A into a new set, we get a set of sets. We call the set of all subsets of A the **power set** of A, and write it $\mathcal{P}(A)$.

Example B.2.4
Let $A = \{1, 2, 3\}$. Find $\mathcal{P}(A)$.

Solution. $\mathcal{P}(A)$ is a set of sets, all of which are subsets of A. So

$$\mathcal{P}(A) = \{\emptyset, \{1\}, \{2\}, \{3\}, \{1,2\}, \{1,3\}, \{2,3\}, \{1,2,3\}\}.$$

Notice that while $2 \in A$, it is wrong to write $2 \in \mathcal{P}(A)$ since none of the elements in $\mathcal{P}(A)$ are numbers! On the other hand, we do have $\{2\} \in \mathcal{P}(A)$ because $\{2\} \subseteq A$.

What does a subset of $\mathcal{P}(A)$ look like? Notice that $\{2\} \not\subseteq \mathcal{P}(A)$ because not everything in $\{2\}$ is in $\mathcal{P}(A)$. But we do have $\{\{2\}\} \subseteq \mathcal{P}(A)$. The only element of $\{\{2\}\}$ is the set $\{2\}$ which is also an element of $\mathcal{P}(A)$. We could take the collection of all subsets of $\mathcal{P}(A)$ and call that $\mathcal{P}(\mathcal{P}(A))$. Or even the power set of that set of sets of sets.

Another way to compare sets is by their *size*. Notice that in the example above, A has 6 elements and B, C, and D all have 3 elements. The size of a set is called the set's **cardinality**. We would write $|A| = 6$, $|B| = 3$, and so on. For sets that have a finite number of elements, the cardinality of the set is simply the number of elements in the set. Note that the cardinality of $\{1,2,3,2,1\}$ is 3. We do not count repeats (in fact, $\{1,2,3,2,1\}$ is exactly the same set as $\{1,2,3\}$). There are sets with infinite cardinality, such as $\mathbb{N}$, the set of rational numbers (written $\mathbb{Q}$), the set of even natural numbers, and the set of real numbers ($\mathbb{R}$). It is possible to distinguish between different infinite cardinalities, but that is beyond the scope of this text. For us, a set will either be infinite, or finite; if it is finite, the we can determine its cardinality by counting elements.

Example B.2.5

1. Find the cardinality of $A = \{23, 24, \ldots, 37, 38\}$.

2. Find the cardinality of $B = \{1, \{2,3,4\}, \emptyset\}$.

3. If $C = \{1,2,3\}$, what is the cardinality of $\mathcal{P}(C)$?

Solution.

1. Since $38 - 23 = 15$, we can conclude that the cardinality of the set is $|A| = 16$ (you need to add one since 23 is included).

2. Here $|B| = 3$. The three elements are the number 1, the set $\{2,3,4\}$, and the empty set.

3. We wrote out the elements of the power set $\mathcal{P}(C)$ above, and there are 8 elements (each of which is a set). So $|\mathcal{P}(C)| = 8$. (You might wonder if there is a relationship between $|A|$ and $|\mathcal{P}(A)|$ for all sets A. This is a good question which we consider in to in Chapter 1.)

B.2.3 Operations On Sets

Is it possible to add two sets? Not really, however there is something similar. If we want to combine two sets to get the collection of objects that are in either set, then we can take the **union** of the two sets. Symbolically,

$$C = A \cup B,$$

read, "C is the union of A and B," means that the elements of C are exactly the elements which are either an element of A or an element of B (or an element of both). For example, if $A = \{1, 2, 3\}$ and $B = \{2, 3, 4\}$, then $A \cup B = \{1, 2, 3, 4\}$.

The other common operation on sets is **intersection**. We write,

$$C = A \cap B$$

and say, "C is the intersection of A and B," when the elements in C are precisely those both in A and in B. So if $A = \{1, 2, 3\}$ and $B = \{2, 3, 4\}$, then $A \cap B = \{2, 3\}$.

Often when dealing with sets, we will have some understanding as to what "everything" is. Perhaps we are only concerned with natural numbers. In this case we would say that our **universe** is $\mathbb{N}$. Sometimes we denote this universe by $\mathcal{U}$. Given this context, we might wish to speak of all the elements which are *not* in a particular set. We say B is the **complement** of A, and write,

$$B = \bar{A}$$

when B contains every element not contained in A. So, if our universe is $\{1, 2, \ldots, 9, 10\}$, and $A = \{2, 3, 5, 7\}$, then $\bar{A} = \{1, 4, 6, 8, 9, 10\}$.

Of course we can perform more than one operation at a time. For example, consider

$$A \cap \bar{B}.$$

This is the set of all elements which are both elements of A and not elements of B. What have we done? We've started with A and removed all of the elements which were in B. Another way to write this is the **set difference**:

$$A \cap \bar{B} = A \setminus B.$$

It is important to remember that these operations (union, intersection, complement, and difference) on sets produce other sets. Don't confuse these with the symbols from the previous section (element of and subset of). $A \cap B$ is a set, while $A \subseteq B$ is true or false. This is the same difference as between $3 + 2$ (which is a number) and $3 \leq 2$ (which is false).

Example B.2.6
Let $A = \{1, 2, 3, 4, 5, 6\}$, $B = \{2, 4, 6\}$, $C = \{1, 2, 3\}$ and $D = \{7, 8, 9\}$. If the universe is $\mathcal{U} = \{1, 2, \ldots, 10\}$, find:

1. $A \cup B$.
2. $A \cap B$.
3. $B \cap C$.
4. $A \cap D$.
5. $\bar{B} \cup C$.
6. $A \setminus B$.
7. $(D \cap \bar{C}) \cup A \cap \bar{B}$.
8. $\emptyset \cup C$.
9. $\emptyset \cap C$.

Solution.

1. $A \cup B = \{1, 2, 3, 4, 5, 6\} = A$ since everything in B is already in A.

2. $A \cap B = \{2, 4, 6\} = B$ since everything in B is in A.

3. $B \cap C = \{2\}$ as the only element of both B and C is 2.

4. $A \cap D = \emptyset$ since A and D have no common elements.

5. $\overline{B \cup C} = \{5, 7, 8, 9, 10\}$. First we find that $B \cup C = \{1, 2, 3, 4, 6\}$, then we take everything not in that set.

6. $A \setminus B = \{1, 3, 5\}$ since the elements 1, 3, and 5 are in A but not in B. This is the same as $A \cap \bar{B}$.

7. $(D \cap \bar{C}) \cup \overline{A \cap B} = \{1, 3, 5, 7, 8, 9, 10\}$. The set contains all elements that are either in D but not in C (i.e., $\{7, 8, 9\}$), or not in both A and B (i.e., $\{1, 3, 5, 7, 8, 9, 10\}$).

8. $\emptyset \cup C = C$ since nothing is added by the empty set.

9. $\emptyset \cap C = \emptyset$ since nothing can be both in a set and in the empty set.

Having notation like this is useful. We will often want to add or remove elements from sets, and our notation allows us to do so precisely.

Example B.2.7
If $A = \{1, 2, 3\}$, then we can describe the set we get by adding the number 4 as $A \cup \{4\}$. If we want to express the set we get by removing the number 2 from A we can do so by writing $A \setminus \{2\}$.

Careful though. If you add an element to the set, you get a new set! So you would have $B = A \cup \{4\}$ and then correctly say that B contains 4, but A does not.

You might notice that the symbols for union and intersection slightly resemble the logic symbols for "or" and "and." This is no accident. What does it mean for x to be an element of $A \cup B$? It means that x is an element of A or x is an element of B (or both). That is,

$$x \in A \cup B \quad \Leftrightarrow \quad x \in A \vee x \in B.$$

Similarly,
$$x \in A \cap B \quad \Leftrightarrow \quad x \in A \wedge x \in B.$$

Also,
$$x \in \bar{A} \quad \Leftrightarrow \quad \neg(x \in A).$$

which says x is an element of the complement of A if x is not an element of A.

There is one more way to combine sets which will be useful for us: the **Cartesian product**, $A \times B$. This sounds fancy but is nothing you haven't seen before. When you graph a function in calculus, you graph it in the Cartesian plane. This is the set of all ordered pairs of real numbers (x, y). We can do this for *any* pair of sets, not just the real numbers with themselves.

Put another way, $A \times B = \{(a, b) : a \in A \wedge b \in B\}$. The first coordinate comes from the first set and the second coordinate comes from the second set. Sometimes we will want to take the Cartesian product of a set with itself, and this is fine: $A \times A = \{(a, b) : a, b \in A\}$ (we might also write A^2 for this set). Notice that in $A \times A$, we still want *all* ordered pairs, not just the ones where the first and second coordinate are the same. We can also take products of 3 or more sets, getting ordered triples, or quadruples, and so on.

> **Example B.2.8**
> Let $A = \{1, 2\}$ and $B = \{3, 4, 5\}$. Find $A \times B$ and $A \times A$. How many elements do you expect to be in $B \times B$?
> **Solution.** $A \times B = \{(1, 3), (1, 4), (1, 5), (2, 3), (2, 4), (2, 5)\}$.
> $A \times A = A^2 = \{(1, 1), (1, 2), (2, 1), (2, 2)\}$.
> $|B \times B| = 9$. There will be 3 pairs with first coordinate 3, three more with first coordinate 4, and a final three with first coordinate 5.

B.2.4 Venn Diagrams

There is a very nice visual tool we can use to represent operations on sets. A **Venn diagram** displays sets as intersecting circles. We can shade the region we are talking about when we carry out an operation. We can also represent cardinality of a particular set by putting the number in the corresponding region.

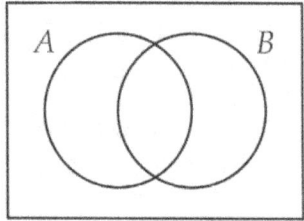 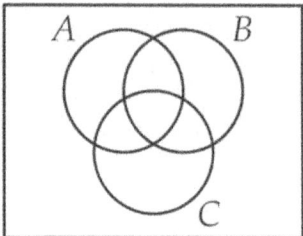

Each circle represents a set. The rectangle containing the circles represents the universe. To represent combinations of these sets, we shade the corresponding region. For example, we could draw $A \cap B$ as:

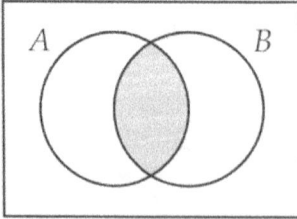

Here is a representation of $A \cap \bar{B}$, or equivalently $A \setminus B$:

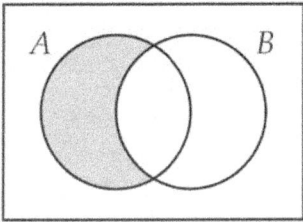

A more complicated example is $(B \cap C) \cup (C \cap \bar{A})$, as seen below.

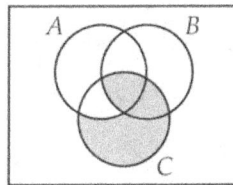

Notice that the shaded regions above could also be arrived at in another way. We could have started with all of C, then excluded the region where C and A overlap outside of B. That region is $(A \cap C) \cap \bar{B}$. So the above Venn diagram also represents $C \cap \overline{((A \cap C) \cap \bar{B})}$. So using just the picture, we have determined that

$$(B \cap C) \cup (C \cap \bar{A}) = C \cap \overline{((A \cap C) \cap \bar{B})}.$$

B.2.5 Exercises

1. Let $A = \{1, 4, 9\}$ and $B = \{1, 3, 6, 10\}$. Find each of the following sets.

    (a) $A \cup B$.

    (b) $A \cap B$.

    (c) $A \setminus B$.

    (d) $B \setminus A$.

    [Solution]

2. Find the least element of each of the following sets, if there is one.

    (a) $\{n \in \mathbb{N} : n^2 - 3 \geq 2\}$.

    (b) $\{n \geq 0 : n^2 - 5 \in \mathbb{N}\}$.

    (c) $\{n^2 + 1 : n \in \mathbb{N}\}$.

    (d) $\{n \in \mathbb{N} : n = k^2 + 1 \text{ for some } k \in \mathbb{N}\}$.

    [Solution]

3. Find the following cardinalities:

    (a) $|A|$ when $A = \{4, 5, 6, \ldots, 37\}$.

    (b) $|A|$ when $A = \{x \in \mathbb{Z} : -2 \leq x \leq 100\}$.

    (c) $|A \cap B|$ when $A = \{x \in \mathbb{N} : x \leq 20\}$ and $B = \{x \in \mathbb{N} : x \text{ is prime}\}$.

    [Solution]

4. Find a set of largest possible size that is a subset of both $\{1,2,3,4,5\}$ and $\{2,4,6,8,10\}$.
   [Solution]

5. Find a set of smallest possible size that has both $\{1,2,3,4,5\}$ and $\{2,4,6,8,10\}$ as subsets.
   [Solution]

6. Let $A = \{n \in \mathbb{N} : 20 \leq n < 50\}$ and $B = \{n \in \mathbb{N} : 10 < n \leq 30\}$. Suppose C is a set such that $C \subseteq A$ and $C \subseteq B$. What is the largest possible cardinality of C?
   [Solution]

7. Let $A = \{1,2,3,4,5\}$ and $B = \{2,3,4\}$. How many sets C have the property that $C \subseteq A$ and $B \subseteq C$.
   [Solution]

   Hint. You should be able to write all of them out. Don't forget A and B, which are also candidates for C.

8. Let $A = \{1,2,3,4,5\}$, $B = \{3,4,5,6,7\}$, and $C = \{2,3,5\}$.

   (a) Find $A \cap B$.

   (b) Find $A \cup B$.

   (c) Find $A \setminus B$.

   (d) Find $A \cap \overline{(B \cup C)}$.

   [Solution]

9. Let $A = \{x \in \mathbb{N} : 4 \leq x < 12\}$ and $B = \{x \in \mathbb{N} : x \text{ is even}\}$.

   (a) Find $A \cap B$.

   (b) Find $A \setminus B$.

   [Solution]

10. Let $A = \{x \in \mathbb{N} : 3 \leq x \leq 13\}$, $B = \{x \in \mathbb{N} : x \text{ is even}\}$, and $C = \{x \in \mathbb{N} : x \text{ is odd}\}$.

    (a) Find $A \cap B$.

    (b) Find $A \cup B$.

    (c) Find $B \cap C$.

    (d) Find $B \cup C$.

11. Find an example of sets A and B such that $A \cap B = \{3,5\}$ and $A \cup B = \{2,3,5,7,8\}$.
    [Solution]

12. Find an example of sets A and B such that $A \subseteq B$ and $A \in B$.
    [Solution]

13. Recall $\mathbb{Z} = \{\ldots, -2, -1, 0, 1, 2, \ldots\}$ (the integers). Let $\mathbb{Z}^+ = \{1, 2, 3, \ldots\}$ be the positive integers. Let $2\mathbb{Z}$ be the even integers, $3\mathbb{Z}$ be the multiples of 3, and so on.

    (a) Is $\mathbb{Z}^+ \subseteq 2\mathbb{Z}$? Explain.

    (b) Is $2\mathbb{Z} \subseteq \mathbb{Z}^+$? Explain.

    (c) Find $2\mathbb{Z} \cap 3\mathbb{Z}$. Describe the set in words, and using set notation.

    (d) Express $\{x \in \mathbb{Z} : \exists y \in \mathbb{Z}(x = 2y \lor x = 3y)\}$ as a union or intersection of two sets already described in this problem.

14. Let A_2 be the set of all multiples of 2 except for 2. Let A_3 be the set of all multiples of 3 except for 3. And so on, so that A_n is the set of all multiples of n except for n, for any $n \geq 2$. Describe (in words) the set $A_2 \cup A_3 \cup A_4 \cup \cdots$.

    Hint. It might help to think about what the union $A_2 \cup A_3$ is first. Then think about what numbers are *not* in that union. What will happen when you also include A_5?

15. Draw a Venn diagram to represent each of the following:

    (a) $A \cup \bar{B}$

    (b) $\overline{(A \cup B)}$

    (c) $A \cap (B \cup C)$

    (d) $(A \cap B) \cup C$

    (e) $\bar{A} \cap B \cap \bar{C}$

    (f) $(A \cup B) \setminus C$

16. Describe a set in terms of A and B (using set notation) which has the following Venn diagram:

    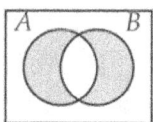

17. Let $A = \{a, b, c, d\}$. Find $\mathcal{P}(A)$.

    Hint. We are looking for a set containing 16 sets.

18. Let $A = \{1, 2, \ldots, 10\}$. How many subsets of A contain exactly one element (i.e., how many singleton subsets are there)?

    How many doubleton subsets (containing exactly two elements) are there?

    Hint. Write these out, or at least start to and look for a pattern.

19. Let $A = \{1, 2, 3, 4, 5, 6\}$. Find all sets $B \in \mathcal{P}(A)$ which have the property $\{2, 3, 5\} \subseteq B$.

20. Find an example of sets A and B such that $|A| = 4$, $|B| = 5$, and $|A \cup B| = 9$.

21. Find an example of sets A and B such that $|A| = 3$, $|B| = 4$, and $|A \cup B| = 5$.

22. Are there sets A and B such that $|A| = |B|$, $|A \cup B| = 10$, and $|A \cap B| = 5$? Explain.

23. Let $A = \{2, 4, 6, 8\}$. Suppose B is a set with $|B| = 5$.

    (a) What are the smallest and largest possible values of $|A \cup B|$? Explain.

    (b) What are the smallest and largest possible values of $|A \cap B|$? Explain.

    (c) What are the smallest and largest possible values of $|A \times B|$? Explain.

24. Let $X = \{n \in \mathbb{N} : 10 \leq n < 20\}$. Find examples of sets with the properties below and very briefly explain why your examples work.

    (a) A set $A \subseteq \mathbb{N}$ with $|A| = 10$ such that $X \setminus A = \{10, 12, 14\}$.

    (b) A set $B \in \mathcal{P}(X)$ with $|B| = 5$.

    (c) A set $C \subseteq \mathcal{P}(X)$ with $|C| = 5$.

    (d) A set $D \subseteq X \times X$ with $|D| = 5$

    (e) A set $E \subseteq X$ such that $|E| \in E$.

25. Let A, B and C be sets.

    (a) Suppose that $A \subseteq B$ and $B \subseteq C$. Does this mean that $A \subseteq C$? Prove your answer. Hint: to prove that $A \subseteq C$ you must prove the implication, "for all x, if $x \in A$ then $x \in C$."

    (b) Suppose that $A \in B$ and $B \in C$. Does this mean that $A \in C$? Give an example to prove that this does NOT always happen (and explain why your example works). You should be able to give an example where $|A| = |B| = |C| = 2$.

26. In a regular deck of playing cards there are 26 red cards and 12 face cards. Explain, using sets and what you have learned about cardinalities, why there are only 32 cards which are either red or a face card.

27. Find an example of a set A with $|A| = 3$ which contains only other sets and has the following property: for all sets $B \in A$, we also have $B \subseteq A$. Explain why your example works. (FYI: sets that have this property are called **transitive**.)

28. Consider the sets A and B, where $A = \{3, |B|\}$ and $B = \{1, |A|, |B|\}$. What are the sets? [Solution]

29. Explain why there is no set A which satisfies $A = \{2, |A|\}$.

    Hint. It looks like you should be able to define the set A like this. But consider the two possible values for $|A|$.

30. Find all sets $A, B,$ and C which satisfy the following.

$$A = \{1, |B|, |C|\}$$
$$B = \{2, |A|, |C|\}$$
$$C = \{1, 2, |A|, |B|\}.$$

B.3 Functions

A **function** is a rule that assigns each input exactly one output. We call the output the **image** of the input. The set of all inputs for a function is called the **domain**. The set of all allowable outputs is called the **codomain**. We would write $f : X \to Y$ to describe a function with name f, domain X and codomain Y. This does not tell us *which* function f is though. To define the function, we must describe the rule. This is often done by giving a formula to compute the output for any input (although this is certainly not the only way to describe the rule).

For example, consider the function $f : \mathbb{N} \to \mathbb{N}$ defined by $f(x) = x^2 + 3$. Here the domain and codomain are the same set (the natural numbers). The rule is: take your input, multiply it by itself and add 3. This works because we can apply this rule to every natural number (every element of the domain) and the result is always a natural number (an element of the codomain). Notice though that not every natural number is actually an output (there is no way to get 0, 1, 2, 5, etc.). The set of natural numbers that *are* outputs is called the **range** of the function (in this case, the range is $\{3, 4, 7, 12, 19, 28, \ldots\}$, all the natural numbers that are 3 more than a perfect square).

The key thing that makes a rule a *function* is that there is *exactly one* output for each input. That is, it is important that the rule be a good rule. What output do we assign to the input 7? There can only be one answer for any particular function.

Example B.3.1

The following are all examples of functions:

1. $f : \mathbb{Z} \to \mathbb{Z}$ defined by $f(n) = 3n$. The domain and codomain are both the set of integers. However, the range is only the set of integer multiples of 3.

2. $g : \{1, 2, 3\} \to \{a, b, c\}$ defined by $g(1) = c$, $g(2) = a$ and $g(3) = a$. The domain is the set $\{1, 2, 3\}$, the codomain is the set $\{a, b, c\}$ and the range is the set $\{a, c\}$. Note that $g(2)$ and $g(3)$ are the same element of the codomain. This is okay since each element in the domain still has only one output.

3. $h : \{1, 2, 3, 4\} \to \mathbb{N}$ defined by the table:

x	1	2	3	4
$h(x)$	3	6	9	12

   Here the domain is the finite set $\{1, 2, 3, 4\}$ and to codomain is the set of natural numbers, $\mathbb{N}$. At first you might think this function is the same as f defined above. It is absolutely not. Even though the rule is the same, the domain and codomain are different, so these are two different functions.

Example B.3.2

Just because you can describe a rule in the same way you would write a function, does not mean that the rule is a function. The following are NOT functions.

1. $f : \mathbb{N} \to \mathbb{N}$ defined by $f(n) = \frac{n}{2}$. The reason this is not a function is because not every input has an output. Where does f send 3? The rule says that $f(3) = \frac{3}{2}$, but $\frac{3}{2}$ is not an element of the codomain.

2. Consider the rule that matches each person to their phone number. If you think of the set of people as the domain and the set of phone numbers as the codomain, then this is not a function, since some people have two phone numbers. Switching the domain and codomain sets doesn't help either, since some phone numbers belong to multiple people (assuming some households still have landlines when you are reading this).

B.3.1 Describing Functions

It is worth making a distinction between a function and its description. The function is the abstract mathematical object that in some way exists whether or not anyone ever talks about it. But when we *do* want to talk about the function, we need a way to describe it. A particular function can be described in multiple ways.

Some calculus textbooks talk about the *Rule of Four*, that every function can be described in four ways: algebraically (a formula), numerically (a table), graphically, or in words. In discrete math, we can still use any of these to describe functions, but we can also be more specific since we are primarily concerned with functions that have $\mathbb{N}$ or a finite subset of $\mathbb{N}$ as their domain.

Describing a function graphically usually means drawing the graph of the function: plotting the points on the plain. We can do this, and might get a graph like the following for a function $f : \{1, 2, 3\} \to \{1, 2, 3\}$.

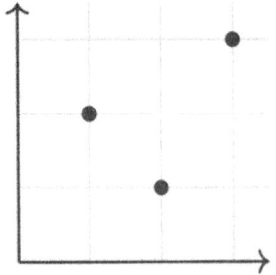

It would be absolutely WRONG to connect the dots or try to fit them to some curve. There are only three elements in the domain. A curve would mean that the domain contains an entire interval of real numbers.

Here is another way to represent that same function:

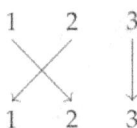

This shows that the function f sends 1 to 2, 2 to 1 and 3 to 3: just follow the arrows.

The arrow diagram used to define the function above can be very helpful in visualizing functions. We will often be working with functions with *finite* domains, so this kind of picture is often more useful than a traditional graph of a function.

Note that for finite domains, finding an algebraic formula that gives the output for any input is often impossible. Of course we could use a piecewise defined function, like

$$f(x) = \begin{cases} x+1 & \text{if } x = 1 \\ x-1 & \text{if } x = 2 \\ x & \text{if } x = 3 \end{cases}.$$

This describes exactly the same function as above, but we can all agree is a ridiculous way of doing so.

Since we will so often use functions with small domains and codomains, let's adopt some notation to describe them. All we need is some clear way of denoting the image of each element in the domain. In fact, writing a table of values would work perfectly:

x	0	1	2	3	4
$f(x)$	3	3	2	4	1

We simplify this further by writing this as a matrix with each input directly over its output:

$$f = \begin{pmatrix} 0 & 1 & 2 & 3 & 4 \\ 3 & 3 & 2 & 4 & 1 \end{pmatrix}.$$

Note this is just notation and not the same sort of matrix you would find in a linear algebra class (it does not make sense to do operations with these matrices, or row reduce them, for example).

One advantage of the two-line notation over the arrow diagrams is that it is harder to accidentally define a rule that is not a function using two-line notation.

Example B.3.3
Which of the following diagrams represent a function? Let $X = \{1, 2, 3, 4\}$ and $Y = \{a, b, c, d\}$.

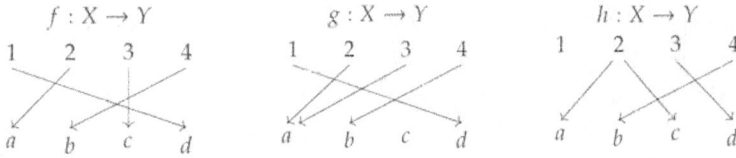

Solution. f is a function. So is g. There is no problem with an element of the codomain not being the image of any input, and there is no problem with a from the codomain being the image of both 2 and 3 from the domain. We could use our two-line notation to write these as

$$f = \begin{pmatrix} 1 & 2 & 3 & 4 \\ d & a & c & b \end{pmatrix} \qquad g = \begin{pmatrix} 1 & 2 & 3 & 4 \\ d & a & a & b \end{pmatrix}.$$

However, h is NOT a function. In fact, it fails for two reasons. First, the element 1 from the domain has not been mapped to any element from the codomain. Second, the element 2 from the domain has been mapped to more than one element from the codomain (a and c). Note that either one of these problems is enough to make a rule not a function. In general, neither of the following mappings are functions:

It might also be helpful to think about how you would write the two-line notation for h. We would have something like:

$$h = \begin{pmatrix} 1 & 2 & 3 & 4 \\ & a,c? & d & b \end{pmatrix}.$$

There is nothing under 1 (bad) and we needed to put more than one thing under 2 (very bad). With a rule that is actually a function, the two-line notation will always "work".

We will also be interested in functions with domain $\mathbb{N}$. Here two-line notation is no good, but describing the function algebraically is often possible. Even tables are a little awkward, since they do not describe the function completely. For example, consider the function $f : \mathbb{N} \to \mathbb{N}$ given by the table below.

x	0	1	2	3	4	5	...
$f(x)$	0	1	4	9	16	25	...

Have I given you enough entries for you to be able to determine $f(6)$? You might guess that $f(6) = 36$, but there is no way for you to *know* this for sure. Maybe I am being a jerk and intended $f(6) = 42$. In fact, for every natural number n, there is a function that agrees with the table above, but for which $f(6) = n$.

Okay, suppose I really did mean for $f(6) = 36$, and in fact, for the rule that you think is governing the function to actually be the rule. Then I should say what that rule is. $f(n) = n^2$. Now there is no confusion possible.

Giving an explicit formula that calculates the image of any element in the domain is a great way to describe a function. We will say that these explicit rules are **closed formulas** for the function.

There is another very useful way to describe functions whose domain is $\mathbb{N}$, that rely specifically on the structure of the natural numbers. We can define a function *recursively*!

Example B.3.4

Consider the function $f : \mathbb{N} \to \mathbb{N}$ given by $f(0) = 0$ and $f(n + 1) = f(n) + 2n + 1$. Find $f(6)$.

Solution. The rule says that $f(6) = f(5) + 11$ (we are using $6 = n + 1$ so $n = 5$). We don't know what $f(5)$ is though. Well, we know that $f(5) = f(4) + 9$. So we need to compute $f(4)$, which will require knowing $f(3)$, which will require $f(2)$,... will it ever end?

Yes! In fact, this process will always end because we have $\mathbb{N}$ as our domain, so there is a least element. And we gave the value of $f(0)$ explicitly, so we are good. In fact, we might decide to work up to $f(6)$ instead of working down from $f(6)$:

$$f(1) = f(0) + 1 = \qquad 0 + 1 = 1$$
$$f(2) = f(1) + 3 = \qquad 1 + 3 = 4$$
$$f(3) = f(2) + 5 = \qquad 4 + 5 = 9$$
$$f(4) = f(3) + 7 = \qquad 9 + 7 = 16$$
$$f(5) = f(4) + 9 = \qquad 16 + 9 = 25$$
$$f(6) = f(5) + 11 = \qquad 25 + 11 = 36$$

It looks that this recursively defined function is the same as the explicitly defined function $f(n) = n^2$. Is it? Later we will prove that it is.

Recursively defined functions are often easier to create from a "real world" problem, because they describe how the values of the functions are changing. However, this comes with a price. It is harder to calculate the image of a single input, since you need to know the images of other (previous) elements in the domain.

Recursively Defined Functions.

For a function $f : \mathbb{N} \to \mathbb{N}$, a **recursive definition** consists of an **initial condition** together with a **recurrence relation**. The initial condition is the explicitly given value of $f(0)$. The recurrence relation is a formula for $f(n + 1)$ in terms for $f(n)$ (and possibly n itself).

Example B.3.5

Give recursive definitions for the functions described below.

1. $f : \mathbb{N} \to \mathbb{N}$ gives the number of snails in your terrarium n years after you built it, assuming you started with 3 snails and the number of snails doubles each year.

2. $g : \mathbb{N} \to \mathbb{N}$ gives the number of push-ups you do n days after you started your push-ups challenge, assuming you could do 7 push-ups on day 0 and you can do 2 more push-ups each day.

3. $h : \mathbb{N} \to \mathbb{N}$ defined by $f(n) = n!$. Recall that $n! = 1\cdot 2\cdot 3\cdots (n-1)\cdot n$ is the product of all numbers from 1 through n. We also define $0! = 1$.

Solution.

1. The initial condition is $f(0) = 3$. To get $f(n+1)$ we would double the number of snails in the terrarium the previous year, which is given by $f(n)$. Thus $f(n+1) = 2f(n)$. The full recursive definition contains both of these, and would be written,
$$f(0) = 3;\ f(n+1) = 2f(n).$$

2. We are told that on day 0 you can do 7 push-ups, so $f(0) = 7$. The number of push-ups you can do on day $n+1$ is 2 more than the number you can do on day n, which is given by $f(n)$. Thus
$$f(0) = 7;\ f(n+1) = f(n) + 2.$$

3. Here $f(0) = 1$. To get the recurrence relation, think about how you can get $f(n+1) = (n+1)!$ from $f(n) = n!$. If you write out both of these as products, you see that $(n+1)!$ is just like $n!$ except you have one more term in the product, an extra $n+1$. So we have,
$$f(0) = 1;\ f(n+1) = (n+1)\cdot f(n).$$

B.3.2 Surjections, Injections, and Bijections

We now turn to investigating special properties functions might or might not possess.

In the examples above, you may have noticed that sometimes there are elements of the codomain which are not in the range. When this sort of the thing *does not* happen, (that is, when everything in the codomain is in the range) we say the function is **onto** or that the function maps the domain *onto* the codomain. This terminology should make sense: the function puts the domain (entirely) on top of the codomain. The fancy math term for an onto function is a **surjection**, and we say that an onto function is a **surjective** function.

In pictures:

Surjective   Not surjective

Example B.3.6

Which functions are surjective (i.e., onto)?

1. $f : \mathbb{Z} \to \mathbb{Z}$ defined by $f(n) = 3n$.

2. $g : \{1, 2, 3\} \to \{a, b, c\}$ defined by $g = \begin{pmatrix} 1 & 2 & 3 \\ c & a & a \end{pmatrix}$.

3. $h : \{1, 2, 3\} \to \{1, 2, 3\}$ defined as follows:

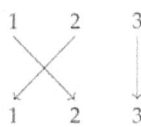

Solution.

1. f is not surjective. There are elements in the codomain which are not in the range. For example, no $n \in \mathbb{Z}$ gets mapped to the number 1 (the rule would say that $\frac{1}{3}$ would be sent to 1, but $\frac{1}{3}$ is not in the domain). In fact, the range of the function is $3\mathbb{Z}$ (the integer multiples of 3), which is not equal to $\mathbb{Z}$.

2. g is not surjective. There is no $x \in \{1, 2, 3\}$ (the domain) for which $g(x) = b$, so b, which is in the codomain, is not in the range. Notice that there is an element from the codomain "missing" from the bottom row of the matrix.

3. h is surjective. Every element of the codomain is also in the range. Nothing in the codomain is missed.

To be a function, a rule cannot assign a single element of the domain to two or more different elements of the codomain. However, we have seen that the reverse *is* permissible: a function might assign the same element of the codomain to two or more different elements of the domain. When this *does not* occur (that is, when each element of the codomain is the image of at most one element of the domain) then we say the function is **one-to-one**. Again, this terminology makes sense: we are sending at most one element from the domain to one element from the codomain. One input to one output. The fancy math term for a one-to-one function is an **injection**. We call one-to-one functions **injective** functions.

In pictures:

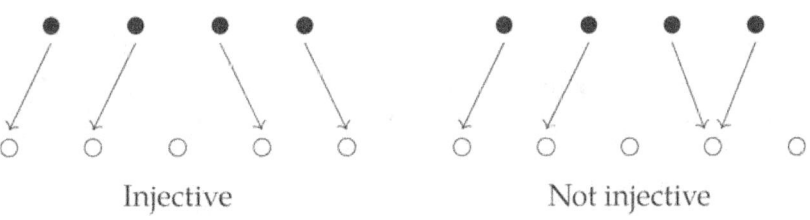

Injective   Not injective

Example B.3.7

Which functions are injective (i.e., one-to-one)?

1. $f : \mathbb{Z} \to \mathbb{Z}$ defined by $f(n) = 3n$.

2. $g : \{1,2,3\} \to \{a,b,c\}$ defined by $g = \begin{pmatrix} 1 & 2 & 3 \\ c & a & a \end{pmatrix}$.

3. $h : \{1,2,3\} \to \{1,2,3\}$ defined as follows:

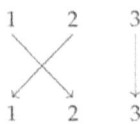

Solution.

1. f is injective. Each element in the codomain is assigned to at *most* one element from the domain. If x is a multiple of three, then only $x/3$ is mapped to x. If x is not a multiple of 3, then there is no input corresponding to the output x.

2. g is not injective. Both inputs 2 and 3 are assigned the output a. Notice that there is an element from the codomain that appears more than once on the bottom row of the matrix.

3. h is injective. Each output is only an output once.

Be careful: "surjective" and "injective" are NOT opposites. You can see in the two examples above that there are functions which are surjective but not injective, injective but not surjective, both, or neither. In the case when a function is both one-to-one and onto (an injection and surjection), we say the function is a **bijection**, or that the function is a **bijective** function.

To illustrate the contrast between these two properties, consider a more formal definition of each, side by side.

Injective vs Surjective.

A function is **injective** provided every element of the codomain is the image of *at most* one element from the domain.

A function is **surjective** provided every element of the codomain is the image of *at least* one element from the domain.

Notice both properties are determined by what happens to elements of the codomain: they could be repeated as images or they could be "missed" (not be images). Injective functions do not have repeats but might or might not miss elements. Surjective functions do not miss elements, but might or might not have repeats. The bijective functions are those that do not have repeats and do not miss elements.

B.3.3 Image and Inverse Image

When discussing functions, we have notation for talking about an element of the domain (say x) and its corresponding element in the codomain (we write $f(x)$, which *is* the image of x). Sometimes we will want to talk about all the elements that are images of some subset of the domain. It would also be nice to start with some element of the codomain (say y) and talk about which element or elements (if any) from the domain it is the image of. We could write "those x in the domain such that $f(x) = y$," but this is a lot of writing. Here is some notation to make our lives easier.

To address the first situation, what we are after is a way to describe the *set* of images of elements in some subset of the domain. Suppose $f : X \to Y$ is a function and that $A \subseteq X$ is some subset of the domain (possibly all of it). We will use the notation $f(A)$ to denote the **image of A under** f, namely the set of elements in Y that are the image of elements from A. That is, $f(A) = \{f(a) \in Y : a \in A\}$.

We can do this in the other direction as well. We might ask which elements of the domain get mapped to a particular set in the codomain. Let $f : X \to Y$ be a function and suppose $B \subseteq Y$ is a subset of the codomain. Then we will write $f^{-1}(B)$ for the **inverse image of B under** f, namely the set of elements in X whose image are elements in B. In other words, $f^{-1}(B) = \{x \in X : f(x) \in B\}$.

Often we are interested in the element(s) whose image is a particular element y of in the codomain. The notation above works: $f^{-1}(\{y\})$ is the set of all elements in the domain that f sends to y. It makes sense to think of this as a set: there might not be anything sent to y (if y is not in the range), in which case $f^{-1}(\{y\}) = \emptyset$. Or f might send multiple elements to y (if f is not injective). As a notational convenience, we usually drop the set braces around the y and write $f \operatorname{inf}(y)$ instead for this set.

WARNING: $f^{-1}(y)$ is not an inverse function! Inverse functions only exist for bijections, but $f^{-1}(y)$ is defined for any function f. The point: $f^{-1}(y)$ is a *set*, not an *element* of the domain. This is just sloppy notation for $f^{-1}(\{y\})$. To help make this distinction, we would call $f^{-1}(y)$ the **complete inverse image of y under** f. It is not the image of y under f^{-1} (since the function f^{-1} might not exist).

Example B.3.8

Consider the function $f : \{1,2,3,4,5,6\} \to \{a,b,c,d\}$ given by

$$f = \begin{pmatrix} 1 & 2 & 3 & 4 & 5 & 6 \\ a & a & b & b & b & c \end{pmatrix}.$$

Find $f(\{1,2,3\})$, $f \operatorname{inf}(\{a,b\})$, and $f^{-1}(d)$.

Solution. $f(\{1,2,3\}) = \{a,b\}$ since a and b are the elements in the codomain to which f sends 1 and 2.

$f \operatorname{inf}(\{a,b\}) = \{1,2,3,4,5\}$ since these are exactly the elements that f sends to a and b.

$f \operatorname{inf}(d) = \emptyset$ since d is not in the range of f.

Example B.3.9

Consider the function $g : \mathbb{Z} \to \mathbb{Z}$ defined by $g(n) = n^2 + 1$. Find $g(1)$ and $g(\{1\})$. Then find $g^{-1}(1)$, $g^{-1}(2)$, and $g^{-1}(3)$.

Solution. Note that $g(1) \neq g(\{1\})$. The first is an element: $g(1) = 2$. The second is a set: $g(\{1\}) = \{2\}$.

To find $g^{-1}(1)$, we need to find all integers n such that $n^2 + 1 = 1$. Clearly only 0 works, so $g^{-1}(1) = \{0\}$ (note that even though there is only one element, we still write it as a set with one element in it).

To find $g^{-1}(2)$, we need to find all n such that $n^2 + 1 = 2$. We see $g^{-1}(2) = \{-1, 1\}$.

Finally, if $n^2 + 1 = 3$, then we are looking for an n such that $n^2 = 2$. There are no such integers so $g^{-1}(3) = \emptyset$.

Since $f^{-1}(y)$ is a set, it makes sense to ask for $|f^{-1}(y)|$, the number of elements in the domain which map to y.

Example B.3.10
Find a function $f : \{1, 2, 3, 4, 5\} \to \mathbb{N}$ such that $|f^{-1}(7)| = 5$.

Solution. There is only one such function. We need five elements of the domain to map to the number $7 \in \mathbb{N}$. Since there are only five elements in the domain, all of them must map to 7. So

$$f = \begin{pmatrix} 1 & 2 & 3 & 4 & 5 \\ 7 & 7 & 7 & 7 & 7 \end{pmatrix}.$$

FUNCTION DEFINITIONS.

Here is a summary of all the main concepts and definitions we use when working with functions.

- A **function** is a rule that assigns each element of a set, called the **domain**, to exactly one element of a second set, called the **codomain**.

- Notation: $f : X \to Y$ is our way of saying that the function is called f, the domain is the set X, and the codomain is the set Y.

- To specify the rule for a function with small domain, use **two-line notation** by writing a matrix with each output directly below its corresponding input, as in:
$$f = \begin{pmatrix} 1 & 2 & 3 & 4 \\ 2 & 1 & 3 & 1 \end{pmatrix}.$$

- $f(x) = y$ means the element x of the domain (input) is assigned to the element y of the codomain. We say y is an output. Alternatively, we call y the **image of** x **under** f.

- The **range** is a subset of the codomain. It is the set of all elements which are assigned to at least one element of the domain by the function. That is, the range is the set of all outputs.

- A function is **injective** (an **injection** or **one-to-one**) if every element of the codomain is the image of **at most** one element from the domain.

- A function is **surjective** (a **surjection** or **onto**) if every element of the codomain is the image of **at least** one element from the domain.

- A **bijection** is a function which is both an injection and surjection. In other words, if every element of the codomain is the image of **exactly one** element from the domain.

- The **image** of an element x in the domain is the element y in the codomain that x is mapped to. That is, the image of x under f is $f(x)$.

- The **complete inverse image** of an element y in the codomain, written $f^{-1}(y)$, is the *set* of all elements in the domain which are assigned to y by the function.

- The **image** of a subset A of the domain is the set $f(A) = \{f(a) \in Y : a \in A\}$.

- The **inverse image** of a a subset B of the codomain is the set $f^{-1}(B) = \{x \in X : f(x) \in B\}$.

B.3.4 Exercises

1. Consider the function $f : \{1,2,3,4\} \to \{1,2,3,4\}$ given by
$$f(n) = \begin{pmatrix} 1 & 2 & 3 & 4 \\ 4 & 1 & 3 & 4 \end{pmatrix}.$$

   (a) Find $f(1)$.

   (b) Find an element n in the domain such that $f(n) = 1$.

   (c) Find an element n of the domain such that $f(n) = n$.

   (d) Find an element of the codomain that is not in the range.

   [Solution]

2. The following functions all have $\{1,2,3,4,5\}$ as both their domain and codomain. For each, determine whether it is (only) injective, (only) surjective, bijective, or neither injective nor surjective.

   (a) $f = \begin{pmatrix} 1 & 2 & 3 & 4 & 5 \\ 3 & 3 & 3 & 3 & 3 \end{pmatrix}.$

   (b) $f = \begin{pmatrix} 1 & 2 & 3 & 4 & 5 \\ 2 & 3 & 1 & 5 & 4 \end{pmatrix}.$

   (c) $f(x) = 6 - x.$

   (d) $f(x) = \begin{cases} x/2 & \text{if } x \text{ is even} \\ (x+1)/2 & \text{if } x \text{ is odd} \end{cases}.$

   [Solution]

3. The following functions all have domain $\{1,2,3,4,5\}$ and codomain $\{1,2,3\}$. For each, determine whether it is (only) injective, (only) surjective, bijective, or neither injective nor surjective.

   (a) $f = \begin{pmatrix} 1 & 2 & 3 & 4 & 5 \\ 1 & 2 & 1 & 2 & 1 \end{pmatrix}.$

   (b) $f = \begin{pmatrix} 1 & 2 & 3 & 4 & 5 \\ 1 & 2 & 3 & 1 & 2 \end{pmatrix}.$

   (c) $f(x) = \begin{cases} x & \text{if } x \leq 3 \\ x - 3 & \text{if } x > 3 \end{cases}.$

4. The following functions all have domain $\{1,2,3,4\}$ and codomain $\{1,2,3,4,5\}$. For each, determine whether it is (only) injective, (only) surjective, bijective, or neither injective nor surjective.

   (a) $f = \begin{pmatrix} 1 & 2 & 3 & 4 \\ 1 & 2 & 5 & 4 \end{pmatrix}.$

   (b) $f = \begin{pmatrix} 1 & 2 & 3 & 4 \\ 1 & 2 & 3 & 2 \end{pmatrix}.$

   (c) $f(x)$ gives the number of letters in the English word for the number x. For example, $f(1) = 3$ since "one" contains three letters.

5. Write out all functions $f : \{1,2,3\} \to \{a,b\}$ (using two-line notation).
   How many functions are there?
   How many are injective?
   How many are surjective?
   How many are bijective?   [Solution]

6. Write out all functions $f : \{1,2\} \to \{a,b,c\}$ (in two-line notation).
   How many functions are there?
   How many are injective?
   How many are surjective?
   How many are bijective?

7. Consider the function $f : \{1,2,3,4,5\} \to \{1,2,3,4\}$ given by the table below:

x	1	2	3	4	5
$f(x)$	3	2	4	1	2

   (a) Is f injective? Explain.

   (b) Is f surjective? Explain.

   (c) Write the function using two-line notation.

   [Solution]

8. Consider the function $f : \{1,2,3,4\} \to \{1,2,3,4\}$ given by the graph below.

```
f(x)
 4          •
 3     •         •
 2
 1           •
    1  2  3  4   x
```

(a) Is f injective? Explain.

(b) Is f surjective? Explain.

(c) Write the function using two-line notation.

9. Consider the function $f : \mathbb{N} \to \mathbb{N}$ given *recursively* by $f(0) = 1$ and $f(n+1) = 2 \cdot f(n)$. Find $f(10)$. [Solution]

10. Suppose $f : \mathbb{N} \to \mathbb{N}$ satisfies the recurrence $f(n+1) = f(n) + 3$. Note that this is not enough information to define the function, since we don't have an initial condition. For each of the initial conditions below, find the value of $f(5)$.

   (a) $f(0) = 0$.

   (b) $f(0) = 1$.

   (c) $f(0) = 2$.

   (d) $f(0) = 100$.

   [Solution]

11. Suppose $f : \mathbb{N} \to \mathbb{N}$ satisfies the recurrence relation

$$f(n+1) = \begin{cases} \frac{f(n)}{2} & \text{if } f(n) \text{ is even} \\ 3f(n) + 1 & \text{if } f(n) \text{ is odd} \end{cases}.$$

Note that with the initial condition $f(0) = 1$, the values of the function are: $f(1) = 4$, $f(2) = 2$, $f(3) = 1$, $f(4) = 4$, and so on, the images cycling through those three numbers. Thus f is NOT injective (and also certainly not surjective). Might it be under other initial conditions?[3]

   (a) If f satisfies the initial condition $f(0) = 5$, is f injective? Explain why or give a specific example of two elements from the domain with the same image.

   (b) If f satisfies the initial condition $f(0) = 3$, is f injective? Explain why or give a specific example of two elements from the domain with the same image.

   (c) If f satisfies the initial condition $f(0) = 27$, then it turns out that $f(105) = 10$ and no two numbers less than 105 have the same image. Could f be injective? Explain.

(d) Prove that no matter what initial condition you choose, the function cannot be surjective.

12. For each function given below, determine whether or not the function is injective and whether or not the function is surjective.

   (a) $f : \mathbb{N} \to \mathbb{N}$ given by $f(n) = n + 4$.

   (b) $f : \mathbb{Z} \to \mathbb{Z}$ given by $f(n) = n + 4$.

   (c) $f : \mathbb{Z} \to \mathbb{Z}$ given by $f(n) = 5n - 8$.

   (d) $f : \mathbb{Z} \to \mathbb{Z}$ given by $f(n) = \begin{cases} n/2 & \text{if } n \text{ is even} \\ (n+1)/2 & \text{if } n \text{ is odd.} \end{cases}$

[Solution]

13. Let $A = \{1, 2, 3, \ldots, 10\}$. Consider the function $f : \mathcal{P}(A) \to \mathbb{N}$ given by $f(B) = |B|$. That is, f takes a subset of A as an input and outputs the cardinality of that set.

   (a) Is f injective? Prove your answer.

   (b) Is f surjective? Prove your answer.

   (c) Find $f^{-1}(1)$.

   (d) Find $f^{-1}(0)$.

   (e) Find $f^{-1}(12)$.

[Solution]

14. Let $X = \{n \in \mathbb{N} : 0 \le n \le 999\}$ be the set of all numbers with three or fewer digits. Define the function $f : X \to \mathbb{N}$ by $f(abc) = a + b + c$, where a, b, and c are the digits of the number in X (write numbers less than 100 with leading 0's to make them three digits). For example, $f(253) = 2+5+3 = 10$.

   (a) Let $A = \{n \in X : 113 \le x \le 122\}$. Find $f(A)$.

   (b) Find $f^{-1}(\{1, 2\})$

   (c) Find $f^{-1}(3)$.

   (d) Find $f^{-1}(28)$.

   (e) Is f injective? Explain.

   (f) Is f surjective? Explain.

15. Consider the set $\mathbb{N}^2 = \mathbb{N} \times \mathbb{N}$, the set of all ordered pairs (a, b) where a and b are natural numbers. Consider a function $f : \mathbb{N}^2 \to \mathbb{N}$ given by $f((a, b)) = a + b$.

   (a) Let $A = \{(a, b) \in \mathbb{N}^2 : a, b \le 10\}$. Find $f(A)$.

   (b) Find $f^{-1}(3)$ and $f^{-1}(\{0, 1, 2, 3\})$.

---

[3]It turns out this is a *really* hard question to answer in general. The *Collatz conjecture* is that no matter what the initial condition is, the function will eventually produce 1 as an output. This is an open problem in mathematics: nobody knows the answer.

(c) Give geometric descriptions of $f^{-1}(n)$ and $f^{-1}(\{0, 1, \ldots, n\})$ for any $n \geq 1$.

(d) Find $|f^{-1}(8)|$ and $|f^{-1}(\{0, 1, \ldots, 8\})|$.

16. Let $f : X \to Y$ be some function. Suppose $3 \in Y$. What can you say about $f^{-1}(3)$ if you know,

    (a) f is injective? Explain.

    (b) f is surjective? Explain.

    (c) f is bijective? Explain.

17. Find a set X and a function $f : X \to \mathbb{N}$ so that $f^{-1}(0) \cup f^{-1}(1) = X$. [Solution]

18. What can you deduce about the sets X and Y if you know,

    (a) there is an injective function $f : X \to Y$? Explain.

    (b) there is a surjective function $f : X \to Y$? Explain.

    (c) there is a bijective function $f : X \to Y$? Explain.

19. Suppose $f : X \to Y$ is a function. Which of the following are possible? Explain.

    (a) f is injective but not surjective.

    (b) f is surjective but not injective.

    (c) $|X| = |Y|$ and f is injective but not surjective.

    (d) $|X| = |Y|$ and f is surjective but not injective.

    (e) $|X| = |Y|$, X and Y are finite, and f is injective but not surjective.

    (f) $|X| = |Y|$, X and Y are finite, and f is surjective but not injective.

20. Let $f : X \to Y$ and $g : Y \to Z$ be functions. We can define the **composition** of f and g to be the function $g \circ f : X \to Z$ for which the image of each $x \in X$ is $g(f(x))$. That is, plug x into f, then plug the result into g (just like composition in algebra and calculus).

    (a) If f and g are both injective, must $g \circ f$ be injective? Explain.

    (b) If f and g are both surjective, must $g \circ f$ be surjective? Explain.

    (c) Suppose $g \circ f$ is injective. What, if anything, can you say about f and g? Explain.

    (d) Suppose $g \circ f$ is surjective. What, if anything, can you say about f and g? Explain.

    Hint. Work with some examples. What if $f = \begin{pmatrix} 1 & 2 & 3 \\ a & a & b \end{pmatrix}$ and $g = \begin{pmatrix} a & b & c \\ 5 & 6 & 7 \end{pmatrix}$?

21. Consider the function $f : \mathbb{Z} \to \mathbb{Z}$ given by $f(n) = \begin{cases} n+1 & \text{if } n \text{ is even} \\ n-3 & \text{if } n \text{ is odd} \end{cases}$

    (a) Is f injective? Prove your answer.

    (b) Is f surjective? Prove your answer.

22. At the end of the semester a teacher assigns letter grades to each of her students. Is this a function? If so, what sets make up the domain and codomain, and is the function injective, surjective, bijective, or neither? [Solution]

23. In the game of *Hearts*, four players are each dealt 13 cards from a deck of 52. Is this a function? If so, what sets make up the domain and codomain, and is the function injective, surjective, bijective, or neither?

24. Seven players are playing 5-card stud. Each player initially receives 5 cards from a deck of 52. Is this a function? If so, what sets make up the domain and codomain, and is the function injective, surjective, bijective, or neither? [Solution]

25. Consider the function $f : \mathbb{N} \to \mathbb{N}$ that gives the number of handshakes that take place in a room of n people assuming everyone shakes hands with everyone else. Give a recursive definition for this function. [Solution]

    Hint. To find the recurrence relation, consider how many *new* handshakes occur when person $n+1$ enters the room.

26. Let $f : X \to Y$ be a function and $A \subseteq X$ be a finite subset of the domain. What can you say about the relationship between $|A|$ and $|f(A)|$? Consider both the general case and what happens when you know f is injective, surjective, or bijective. [Solution]

27. Let $f : X \to Y$ be a function and $B \subseteq Y$ be a finite subset of the codomain. What can you say about the relationship between $|B|$ and $|f^{-1}(B)|$? Consider both the general case and what happens when you know f is injective, surjective, or bijective. [Solution]

28. Let $f : X \to Y$ be a function, $A \subseteq X$ and $B \subseteq Y$.

    (a) Is $f^{-1}(f(A)) = A$? Always, sometimes, never? Explain.

    (b) Is $f(f^{-1}(B)) = B$? Always, sometimes, never? Explain.

    (c) If one or both of the above do not always hold, is there something else you can say? Will equality always hold for particular types of functions? Is there some other relationship other than equality that would always hold? Explore.

29. Let $f : X \to Y$ be a function and $A, B \subseteq X$ be subsets of the domain.

    (a) Is $f(A \cup B) = f(A) \cup f(B)$? Always, sometimes, or never? Explain.

    (b) Is $f(A \cap B) = f(A) \cap f(B)$? Always, sometimes, or never? Explain.

    Hint. One of these is not always true. Try some examples!

30. Let $f : X \to Y$ be a function and $A, B \subseteq Y$ be subsets of the codomain.

   (a) Is $f^{-1}(A \cup B) = f^{-1}(A) \cup f^{-1}(B)$? Always, sometimes, or never? Explain.

   (b) Is $f^{-1}(A \cap B) = f^{-1}(A) \cap f^{-1}(B)$? Always, sometimes, or never? Explain.

B.4 Propositional Logic

Puzzle 334

You stumble upon two trolls playing Stratego®. They tell you:

Troll 1: If we are cousins, then we are both knaves.

Troll 2: We are cousins or we are both knaves.

Could both trolls be knights? Recall that all trolls are either always-truth-telling knights or always-lying knaves.

A **proposition** is simply a statement. **Propositional logic** studies the ways statements can interact with each other. It is important to remember that propositional logic does not really care about the content of the statements. For example, in terms of propositional logic, the claims, "if the moon is made of cheese then basketballs are round," and "if spiders have eight legs then Sam walks with a limp" are exactly the same. They are both implications: statements of the form, $P \to Q$.

B.4.1 Truth Tables

Here's a question about playing Monopoly:

If you get more doubles than any other player then you will lose, or
if you lose then you must have bought the most properties.

True or false? We will answer this question, and won't need to know anything about Monopoly. Instead we will look at the logical *form* of the statement.

We need to decide when the statement $(P \to Q) \vee (Q \to R)$ is true. Using the definitions of the connectives in Section B.1, we see that for this to be true, either $P \to Q$ must be true or $Q \to R$ must be true (or both). Those are true if either P is false or Q is true (in the first case) and Q is false or R is true (in the second case). So—yeah, it gets kind of messy. Luckily, we can make a chart to keep track of all the possibilities. Enter **truth tables**. The idea is this: on each row, we list a possible combination of T's and F's (for true and false) for each of the sentential variables, and then mark down whether the statement in question is true or false in that case. We do this for every possible combination of T's and F's. Then we can clearly see in which cases the statement is true or false. For complicated statements, we will first fill in values for each part of the statement, as a way of breaking up our task into smaller, more manageable pieces.

Since the truth value of a statement is completely determined by the truth values of its parts and how they are connected, all you really need to know is the truth tables for each of the logical connectives. Here they are:

P	Q	$P \wedge Q$	P	Q	$P \vee Q$	P	Q	$P \to Q$	P	Q	$P \leftrightarrow Q$
T	T	T	T	T	T	T	T	T	T	T	T
T	F	F	T	F	T	T	F	F	T	F	F
F	T	F	F	T	T	F	T	T	F	T	F
F	F	F	F	F	F	F	F	T	F	F	T

The truth table for negation looks like this:

P	$\neg P$
T	F
F	T

None of these truth tables should come as a surprise; they are all just restating the definitions of the connectives. Let's try another one.

Example B.4.1

Make a truth table for the statement $\neg P \vee Q$.

Solution. Note that this statement is not $\neg(P \vee Q)$, the negation belongs to P alone. Here is the truth table:

P	Q	$\neg P$	$\neg P \vee Q$
T	T	F	T
T	F	F	F
F	T	T	T
F	F	T	T

We added a column for $\neg P$ to make filling out the last column easier. The entries in the $\neg P$ column were determined by the entries in the P column. Then to fill in the final column, look only at the column for Q and the column for $\neg P$ and use the rule for $\vee$.

Now let's answer our question about monopoly:

Example B.4.2

Analyze the statement, "if you get more doubles than any other player you will lose, or that if you lose you must have bought the most properties," using truth tables.

Solution. Represent the statement in symbols as $(P \to Q) \vee (Q \to R)$, where P is the statement "you get more doubles than any other player," Q is the statement "you will lose," and R is the statement "you must have bought the most properties." Now make a truth table.

The truth table needs to contain 8 rows in order to account for every possible combination of truth and falsity among the three statements. Here is the full truth table:

P	Q	R	$P \to Q$	$Q \to R$	$(P \to Q) \vee (Q \to R)$
T	T	T	T	T	T
T	T	F	T	F	T
T	F	T	F	T	T
T	F	F	F	T	T
F	T	T	T	T	T
F	T	F	T	F	T
F	F	T	T	T	T
F	F	F	T	T	T

The first three columns are simply a systematic listing of all possible combinations of T and F for the three statements (do you see how you would list the 16 possible combinations for four statements?). The next two columns are determined by the values of P, Q, and R and the definition of implication. Then, the last column is determined by the values in the previous two columns and the definition of $\vee$. It is this final column we care about.

Notice that in each of the eight possible cases, the statement in question is true. So our statement about monopoly is true (regardless of how many properties you own, how many doubles you roll, or whether you win or lose).

The statement about monopoly is an example of a **tautology**, a statement which is true on the basis of its logical form alone. Tautologies are always true but they don't tell us much about the world. No knowledge about monopoly was required to determine that the statement was true. In fact, it is equally true that "If the moon is made of cheese, then Elvis is still alive, or if Elvis is still alive, then unicorns have 5 legs."

B.4.2 Logical Equivalence

You might have noticed in Example B.4.1 that the final column in the truth table for $\neg P \vee Q$ is identical to the final column in the truth table for $P \to Q$:

P	Q	$P \to Q$	$\neg P \vee Q$
T	T	T	T
T	F	F	F
F	T	T	T
F	F	T	T

This says that no matter what P and Q are, the statements $\neg P \vee Q$ and $P \to Q$ either both true or both false. We therefore say these statements are **logically equivalent**.

> **Logical Equivalence.**
>
> Two (molecular) statements P and Q are **logically equivalent** provided P is true precisely when Q is true. That is, P and Q have the same truth value under any assignment of truth values to their atomic parts.
>
> To verify that two statements are logically equivalent, you can make a truth table for each and check whether the columns for the two statements are identical.

Recognizing two statements as logically equivalent can be very helpful. Rephrasing a mathematical statement can often lend insight into what it is saying, or how to prove or refute it. By using truth tables we can systematically verify that two statements are indeed logically equivalent.

Example B.4.3

Are the statements, "it will not rain or snow" and "it will not rain and it will not snow" logically equivalent?

Solution. We want to know whether $\neg(P \vee Q)$ is logically equivalent to $\neg P \wedge \neg Q$. Make a truth table which includes both statements:

P	Q	$\neg(P \vee Q)$	$\neg P \wedge \neg Q$
T	T	F	F
T	F	F	F
F	T	F	F
F	F	T	T

Since in every row the truth values for the two statements are equal, the two statements are logically equivalent.

Notice that this example gives us a way to "distribute" a negation over a disjunction (an "or"). We have a similar rule for distributing over conjunctions ("and"s):

De Morgan's Laws.

$\neg(P \wedge Q)$ is logically equivalent to $\neg P \vee \neg Q$.

$\neg(P \vee Q)$ is logically equivalent to $\neg P \wedge \neg Q$.

This suggests there might be a sort of "algebra" you could apply to statements (okay, there is: it is called *Boolean algebra*) to transform one statement into another. We can start collecting useful examples of logical equivalence, and apply them in succession to a statement, instead of writing out a complicated truth table.

De Morgan's laws do not do not directly help us with implications, but as we saw above, every implication can be written as a disjunction:

Implications are Disjuctions.

$P \rightarrow Q$ is logically equivalent to $\neg P \vee Q$.

Example: "If a number is a multiple of 4, then it is even" is equivalent to, "a number is not a multiple of 4 or (else) it is even."

With this and De Morgan's laws, you can take any statement and *simplify* it to the point where negations are only being applied to atomic propositions. Well, actually not, because you could get multiple negations stacked up. But this can be easily dealt with:

Double Negation.

$\neg\neg P$ is logically equivalent to P.

Example: "It is not the case that c is not odd" means "c is odd."

Let's see how we can apply the equivalences we have encountered so far.

> **Example B.4.4**
>
> Prove that the statements $\neg(P \to Q)$ and $P \wedge \neg Q$ are logically equivalent without using truth tables.
>
> **Solution.** We want to start with one of the statements, and transform it into the other through a sequence of logically equivalent statements. Start with $\neg(P \to Q)$. We can rewrite the implication as a disjunction this is logically equivalent to
> $$\neg(\neg P \vee Q).$$
> Now apply DeMorgan's law to get
> $$\neg\neg P \wedge \neg Q.$$
> Finally, use double negation to arrive at $P \wedge \neg Q$

Notice that the above example illustrates that the negation of an implication is NOT an implication: it is a conjunction! We saw this before, in Section B.1, but it is so important and useful, it warants a second blue box here:

> **Negation of an Implication.**
>
> The negation of an implication is a conjuction:
> $$\neg(P \to Q) \text{ is logically equivalent to } P \wedge \neg Q.$$
> That is, the only way for an implication to be false is for the hypothesis to be true *AND* the conclusion to be false.

To verify that two statements are logically equivalent, you can use truth tables or a sequence of logically equivalent replacements. The truth table method, although cumbersome, has the advantage that it can verify that two statements are NOT logically equivalent.

> **Example B.4.5**
>
> Are the statements $(P \vee Q) \to R$ and $(P \to R) \vee (Q \to R)$ logically equivalent?
>
> **Solution.** Note that while we could start rewriting these statements with logically equivalent replacements in the hopes of transforming one into another, we will never be sure that our failure is due to their lack of logical equivalence rather than our lack of imagination. So instead, let's make a truth table:

P	Q	R	$(P \vee Q) \rightarrow R$	$(P \rightarrow R) \vee (Q \rightarrow R)$
T	T	T	T	T
T	T	F	F	F
T	F	T	T	T
T	F	F	F	T
F	T	T	T	T
F	T	F	F	T
F	F	T	T	T
F	F	F	T	T

Look at the fourth (or sixth) row. In this case, $(P \rightarrow R) \vee (Q \rightarrow R)$ is true, but $(P \vee Q) \rightarrow R$ is false. Therefore the statements are not logically equivalent.

While we don't have logical equivalence, it is the case that whenever $(P \vee Q) \rightarrow R$ is true, so is $(P \rightarrow R) \vee (Q \rightarrow R)$. This tells us that we can *deduce* $(P \rightarrow R) \vee (Q \rightarrow R)$ from $(P \vee Q) \rightarrow R$, just not the reverse direction.

B.4.3 Deductions

Puzzle 335

Holmes owns two suits: one black and one tweed. He always wears either a tweed suit or sandals. Whenever he wears his tweed suit and a purple shirt, he chooses to not wear a tie. He never wears the tweed suit unless he is also wearing either a purple shirt or sandals. Whenever he wears sandals, he also wears a purple shirt. Yesterday, Holmes wore a bow tie. What else did he wear?

Earlier we claimed that the following was a valid argument:

> If Edith eats her vegetables, then she can have a cookie. Edith ate her vegetables. Therefore Edith gets a cookie.

How do we know this is valid? Let's look at the form of the statements. Let P denote "Edith eats her vegetables" and Q denote "Edith can have a cookie." The logical form of the argument is then:

$$\begin{array}{r} P \rightarrow Q \\ P \\ \hline \therefore \quad Q \end{array}$$

This is an example of a **deduction rule**, an argument form which is always valid. This one is a particularly famous rule called *modus ponens*. Are you convinced that it is a valid deduction rule? If not, consider the following truth table:

P	Q	$P \to Q$
T	T	T
T	F	F
F	T	T
F	F	T

This is just the truth table for $P \to Q$, but what matters here is that all the lines in the deduction rule have their own column in the truth table. Remember that an argument is valid provided the conclusion must be true given that the premises are true. The premises in this case are $P \to Q$ and P. Which *rows* of the truth table correspond to both of these being true? P is true in the first two rows, and of those, only the first row has $P \to Q$ true as well. And lo-and-behold, in this one case, Q is also true. So if $P \to Q$ and P are both true, we see that Q must be true as well.

Here are a few more examples.

Example B.4.6
Show that

$$P \to Q$$
$$\neg P \to Q$$
$$\therefore Q$$

is a valid deduction rule.

Solution. We make a truth table which contains all the lines of the argument form:

P	Q	$P \to Q$	$\neg P$	$\neg P \to Q$
T	T	T	F	T
T	F	F	F	T
F	T	T	T	T
F	F	T	T	F

(we include a column for $\neg P$ just as a step to help getting the column for $\neg P \to Q$).

Now look at all the rows for which both $P \to Q$ and $\neg P \to Q$ are true. This happens only in rows 1 and 3. Hey! In those rows Q is true as well, so the argument form is valid (it is a valid deduction rule).

Example B.4.7
Decide whether

$$P \to R$$
$$Q \to R$$
$$R$$
$$\therefore P \vee Q$$

is a valid deduction rule.

Solution. Let's make a truth table containing all four statements.

P	Q	R	$P \to R$	$Q \to R$	$P \vee Q$
T	T	T	T	T	T
T	T	F	F	F	T
T	F	T	T	T	T
T	F	F	F	T	T
F	T	T	T	T	T
F	T	F	T	F	T
F	F	T	T	T	F
F	F	F	T	T	F

Look at the second to last row. Here all three premises of the argument are true, but the conclusion is false. Thus this is not a valid deduction rule.

While we have the truth table in front of us, look at rows 1, 3, and 5. These are the only rows in which all of the statements statements $P \to R$, $Q \to R$, and $P \vee Q$ are true. It also happens that R is true in these rows as well. Thus we have discovered a new deduction rule we know *is* valid:

$$P \to R$$
$$Q \to R$$
$$\underline{P \vee Q}$$
$$\therefore \quad R$$

B.4.4 BEYOND PROPOSITIONS

As we saw in Section B.1, not every statement can be analyzed using logical connectives alone. For example, we might want to work with the statement:

All primes greater than 2 are odd.

To write this statement symbolically, we must use quantifiers. We can translate as follows:

$$\forall x((P(x) \wedge x > 2) \to O(x)).$$

In this case, we are using $P(x)$ to denote "x is prime" and $O(x)$ to denote "x is odd." These are not propositions, since their truth value depends on the input x. Better to think of P and O as denoting *properties* of their input. The technical term for these is **predicates** and when we study them in logic, we need to use **predicate logic**.

It is important to stress that predicate logic *extends* propositional logic (much in the way quantum mechanics extends classical mechanics). You will notice that our statement above still used the (propositional) logical connectives. Everything that we learned about logical equivalence and deductions still applies. However, predicate logic allows us to analyze statements at a higher resolution, digging down into the individual propositions P, Q, etc.

A full treatment of predicate logic is beyond the scope of this text. One reason is that there is no systematic procedure for deciding whether two statements in predicate logic are logically equivalent (i.e., there is no analogue to truth tables here). Rather, we end with a two examples of logical equivalence and deduction, to pique your interest.

Example B.4.8

Suppose we claim that there is no smallest number. We can translate this into symbols as
$$\neg \exists x \forall y (x \le y)$$
(literally, "it is not true that there is a number x such that for all numbers y, x is less than or equal to y").

However, we know how negation interacts with quantifiers: we can pass a negation over a quantifier by switching the quantifier type (between universal and existential). So the statement above should be *logically equivalent* to
$$\forall x \exists y (y < x).$$
Notice that $y < x$ is the negation of $x \le y$. This literally says, "for every number x there is a number y which is smaller than x." We see that this is another way to make our original claim.

Example B.4.9

Can you switch the order of quantifiers? For example, consider the two statements:
$$\forall x \exists y P(x, y) \quad \text{and} \quad \exists y \forall x P(x, y).$$
Are these logically equivalent?

Solution. These statements are NOT logically equivalent. To see this, we should provide an interpretation of the predicate $P(x, y)$ which makes one of the statements true and the other false.

Let $P(x, y)$ be the predicate $x < y$. It is true, in the natural numbers, that for all x there is some y greater than that x (since there are infinitely many numbers). However, there is not a natural number y which is greater than every number x. Thus it is possible for $\forall x \exists y P(x, y)$ to be true while $\exists y \forall x P(x, y)$ is false.

We cannot do the reverse of this though. If there is some y for which every x satisfies $P(x, y)$, then certainly for every x there is some y which satisfies $P(x, y)$. The first is saying we can find one y that works for every x. The second allows different y's to work for different x's, but there is nothing preventing us from using the same y that work for every x. In other words, while we don't have logical equivalence between the two statements, we do have a valid deduction rule:

$$\frac{\exists y \forall x P(x, y)}{\therefore \quad \forall x \exists y P(x, y)}$$

Put yet another way, this says that the single statement
$$\exists y \forall x P(x, y) \to \forall x \exists y P(x, y)$$
is always true. This is sort of like a tautology, although we reserve that term for necessary truths in propositional logic. A statement in predicate logic that is necessarily true gets the more prestigious designation of a **law of logic** (or sometimes **logically valid**, but that is less fun).

B.4.5 Exercises

1. Consider the statement about a party, "If it's your birthday or there will be cake, then there will be cake."

    (a) Translate the above statement into symbols. Clearly state which statement is P and which is Q.

    (b) Make a truth table for the statement.

    (c) Assuming the statement is true, what (if anything) can you conclude if there will be cake?

    (d) Assuming the statement is true, what (if anything) can you conclude if there will not be cake?

    (e) Suppose you found out that the statement was a lie. What can you conclude?

    [Solution]

2. Make a truth table for the statement $(P \vee Q) \to (P \wedge Q)$. [Solution]

3. Make a truth table for the statement $\neg P \wedge (Q \to P)$. What can you conclude about P and Q if you know the statement is true? [Solution]

4. Make a truth table for the statement $\neg P \to (Q \wedge R)$.

    Hint. Like above, only now you will need 8 rows instead of just 4.

5. Geoff Poshingten is out at a fancy pizza joint, and decides to order a calzone. When the waiter asks what he would like in it, he replies, "I want either pepperoni or sausage. Also, if I have sausage, then I must also include quail. Oh, and if I have pepperoni or quail then I must also have ricotta cheese."

    (a) Translate Geoff's order into logical symbols.

    (b) The waiter knows that Geoff is either a liar or a truth-teller (so either everything he says is false, or everything is true). Which is it?

    (c) What, if anything, can the waiter conclude about the ingredients in Geoff's desired calzone?

    Hint. You should write down three statements using the symbols P, Q, R, S. If Geoff is a truth-teller, then all three statements would be true. If he was a liar, then all three statements would be false. But in either case, we don't yet know whether the four atomic statements are true or false, since he hasn't said them by themselves.

    A truth table might help, although is probably not entirely necessary.

6. Determine whether the following two statements are logically equivalent: $\neg(P \to Q)$ and $P \wedge \neg Q$. Explain how you know you are correct. [Solution]

7. Are the statements $P \to (Q \vee R)$ and $(P \to Q) \vee (P \to R)$ logically equivalent?

8. Simplify the following statements (so that negation only appears right before variables).

    (a) $\neg(P \to \neg Q)$.

    (b) $(\neg P \vee \neg Q) \to \neg(\neg Q \wedge R)$.

    (c) $\neg((P \to \neg Q) \vee \neg(R \wedge \neg R))$.

    (d) It is false that if Sam is not a man then Chris is a woman, and that Chris is not a woman.

[Solution]

9. Use De Morgan's Laws, and any other logical equivalence facts you know to simplify the following statements. Show all your steps. Your final statements should have negations only appear directly next to the sentence variables or predicates (P, Q, $E(x)$, etc.), and no double negations. It would be a good idea to use only conjunctions, disjunctions, and negations.

    (a) $\neg((\neg P \wedge Q) \vee \neg(R \vee \neg S))$.

    (b) $\neg((\neg P \to \neg Q) \wedge (\neg Q \to R))$ (careful with the implications).

    (c) For both parts above, verify your answers are correct using truth tables. That is, use a truth table to check that the given statement and your proposed simplification are actually logically equivalent.

10. Consider the statement, "If a number is triangular or square, then it is not prime"

    (a) Make a truth table for the statement $(T \vee S) \to \neg P$.

    (b) If you believed the statement was *false*, what properties would a counterexample need to possess? Explain by referencing your truth table.

    (c) If the statement were true, what could you conclude about the number 5657, which is definitely prime? Again, explain using the truth table.

    Hint.

    (a) There will be three rows in which the statement is false.

    (b) Consider the three rows that evaluate to false and say what the truth values of T, S, and P are there.

    (c) You are looking for a row in which P is true, and the whole statement is true.

11. Tommy Flanagan was telling you what he ate yesterday afternoon. He tells you, "I had either popcorn or raisins. Also, if I had cucumber sandwiches, then I had soda. But I didn't drink soda or tea." Of course you know that Tommy is the worlds worst liar, and everything he says is false. What did Tommy eat?

    Justify your answer by writing all of Tommy's statements using sentence variables (P, Q, R, S, T), taking their negations, and using these to deduce what Tommy actually ate.

Hint. Write down three statements, and then take the negation of each (since he is a liar). You should find that Tommy ate one item and drank one item. (Q is for cucumber sandwiches.)

12. Determine if the following deduction rule is valid:
$$\frac{\begin{array}{c} P \vee Q \\ \neg P \end{array}}{\therefore \quad Q}$$

[Solution]

13. Determine if the following is a valid deduction rule:
$$\frac{\begin{array}{c} P \to (Q \vee R) \\ \neg(P \to Q) \end{array}}{\therefore \quad R}$$

14. Determine if the following is a valid deduction rule:
$$\frac{\begin{array}{c} (P \wedge Q) \to R \\ \neg P \vee \neg Q \end{array}}{\therefore \quad \neg R}$$

15. Can you chain implications together? That is, if $P \to Q$ and $Q \to R$, does that means the $P \to R$? Can you chain more implications together? Let's find out:

    (a) Prove that the following is a valid deduction rule:
    $$\frac{\begin{array}{c} P \to Q \\ Q \to R \end{array}}{\therefore \quad P \to R}$$

    (b) Prove that the following is a valid deduction rule for any $n \geq 2$:
    $$\frac{\begin{array}{c} P_1 \to P_2 \\ P_2 \to P_3 \\ \vdots \\ P_{n-1} \to P_n \end{array}}{\therefore \quad P_1 \to P_n.}$$

    I suggest you don't go through the trouble of writing out a 2^n row truth table. Instead, you should use part (a) and mathematical induction.

    Hint. For the second part, you can inductively assume that from the first $n - 2$ implications you can deduce $P_1 \to P_{n-1}$. Then you are back in the case in part (a) again.

16. We can also simplify statements in predicate logic using our rules for passing negations over quantifiers, and then applying propositional logical equivalence to the "inside" propositional part. Simplify the statements below (so negation appears only directly next to predicates).

    (a) $\neg \exists x \forall y (\neg O(x) \vee E(y))$.

    (b) $\neg \forall x \neg \forall y \neg (x < y \wedge \exists z (x < z \vee y < z))$.

    (c) There is a number n for which no other number is either less n than or equal to n.

(d) It is false that for every number n there are two other numbers which n is between.

17. Simplify the statements below to the point that negation symbols occur only directly next to predicates. [Solution]

    (a) $\neg \forall x \forall y (x < y \lor y < x)$.

    (b) $\neg (\exists x P(x) \to \forall y P(y))$.

18. Simplifying negations will be especially useful in the next section when we try to prove a statement by considering what would happen if it were false. For each statement below, write the *negation* of the statement as simply as possible. Don't just say, "it is false that …".

    (a) Every number is either even or odd.

    (b) There is a sequence that is both arithmetic and geometric.

    (c) For all numbers n, if n is prime, then $n + 3$ is not prime.

    Hint. It might help to translate the statements into symbols and then use the formulaic rules to simplify negations (i.e., rules for quantifiers and De Morgan's laws). After simplifying, you should get $\forall x (\neg E(x) \land \neg O(x))$, for the first one, for example. Then translate this back into English.

19. Suppose P and Q are (possibly molecular) propositional statements. Prove that P and Q are logically equivalent if any only if $P \leftrightarrow Q$ is a tautology.

    Hint. What do these concepts mean in terms of truth tables?

20. Suppose $P_1, P_2, \ldots, P_n$ and Q are (possibly molecular) propositional statements. Suppose further that
$$\begin{array}{c} P_1 \\ P_2 \\ \vdots \\ \underline{P_n} \\ \therefore \ Q \end{array}$$
is a valid deduction rule. Prove that the statement
$$(P_1 \land P_2 \land \cdots \land P_n) \to Q$$
is a tautology.

B.5 Proofs

Puzzle 336

Decide which of the following are valid proofs of the following statement:

If ab is an even number, then a or b is even.

1. Suppose a and b are odd. That is, $a = 2k + 1$ and $b = 2m + 1$ for some integers k and m. Then

$$ab = (2k+1)(2m+1)$$
$$= 4km + 2k + 2m + 1$$
$$= 2(2km + k + m) + 1.$$

Therefore ab is odd.

2. Assume that a or b is even - say it is a (the case where b is even will be identical). That is, $a = 2k$ for some integer k. Then

$$ab = (2k)b$$
$$= 2(kb).$$

Thus ab is even.

3. Suppose that ab is even but a and b are both odd. Namely, $ab = 2n$, $a = 2k + 1$ and $b = 2j + 1$ for some integers n, k, and j. Then

$$2n = (2k+1)(2j+1)$$
$$2n = 4kj + 2k + 2j + 1$$
$$n = 2kj + k + j + \frac{1}{2}.$$

But since $2kj + k + j$ is an integer, this says that the integer n is equal to a non-integer, which is impossible.

4. Let ab be an even number, say $ab = 2n$, and a be an odd number, say $a = 2k + 1$.

$$ab = (2k+1)b$$
$$2n = 2kb + b$$
$$2n - 2kb = b$$
$$2(n - kb) = b.$$

Therefore b must be even.

Anyone who doesn't believe there is creativity in mathematics clearly has not tried to write proofs. Finding a way to convince the world that a particular statement is necessarily true is a mighty undertaking and can often be quite challenging. There is not a guaranteed path to success in the search for proofs. For example, in the summer of 1742, a German mathematician by the name of Christian Goldbach wondered whether every even integer greater than 2 could be written as the sum of two primes. Centuries later, we still don't have a proof of this apparent fact (computers have checked that "Goldbach's Conjecture" holds for all numbers less than 4×10^{18}, which leaves only infinitely many more numbers to check).

Writing proofs is a bit of an art. Like any art, to be truly great at it, you need some sort of inspiration, as well as some foundational technique. Just as musicians can learn proper fingering, and painters can learn the proper way to hold a brush, we can look at the proper way to construct arguments. A good place to start might be to study a classic.

Theorem B.5.1
There are infinitely many primes.

Proof. Suppose this were not the case. That is, suppose there are only finitely many primes. Then there must be a last, largest prime, call it p. Consider the number
$$N = p! + 1 = (p \cdot (p-1) \cdots 3 \cdot 2 \cdot 1) + 1.$$

Now N is certainly larger than p. Also, N is not divisible by any number less than or equal to p, since every number less than or equal to p divides $p!$. Thus the prime factorization of N contains prime numbers (possibly just N itself) all greater than p. So p is not the largest prime, a contradiction. Therefore there are infinitely many primes. ∎

This proof is an example of a *proof by contradiction*, one of the standard styles of mathematical proof. First and foremost, the proof is an argument. It contains sequence of statements, the last being the *conclusion* which follows from the previous statements. The argument is valid so the conclusion must be true if the premises are true. Let's go through the proof line by line.

1. Suppose there are only finitely many primes. *[this is a premise. Note the use of "suppose."]*

2. There must be a largest prime, call it p. *[follows from line 1, by the definition of "finitely many."]*

3. Let $N = p! + 1$. *[basically just notation, although this is the inspired part of the proof; looking at $p! + 1$ is the key insight.]*

4. N is larger than p. *[by the definition of $p!$]*

5. N is not divisible by any number less than or equal to p. *[by definition, $p!$ is divisible by each number less than or equal to p, so $p! + 1$ is not.]*

6. The prime factorization of N contains prime numbers greater than p. *[since N is divisible by each prime number in the prime factorization of N, and by line 5.]*

7. Therefore p is not the largest prime. *[by line 6, N is divisible by a prime larger than p.]*

8. This is a contradiction. *[from line 2 and line 7: the largest prime is p and there is a prime larger than p.]*

9. Therefore there are infinitely many primes. *[from line 1 and line 8: our only premise lead to a contradiction, so the premise is false.]*

We should say a bit more about the last line. Up through line 8, we have a valid argument with the premise "there are only finitely many primes" and the conclusion "there is a prime larger than the largest prime." This is a valid argument as each line follows from previous lines. So if the premises are true, then the conclusion *must* be true. However, the conclusion is NOT true. The only way out: the premise must be false.

The sort of line-by-line analysis we did above is a great way to really understand what is going on. Whenever you come across a proof in a textbook, you really should make sure you understand what each line is saying and why it is true. Additionally, it is equally important to understand the overall structure of the proof. This is where using tools from logic is helpful. Luckily there are a relatively small number of standard proof styles that keep showing up again and again. Being familiar with these can help understand proof, as well as give ideas of how to write your own.

B.5.1 Direct Proof

The simplest (from a logic perspective) style of proof is a **direct proof**. Often all that is required to prove something is a systematic explanation of what everything means. Direct proofs are especially useful when proving implications. The general format to prove $P \rightarrow Q$ is this:

Assume P. Explain, explain, ..., explain. Therefore Q.

Often we want to prove universal statements, perhaps of the form $\forall x (P(x) \rightarrow Q(x))$. Again, we will want to assume $P(x)$ is true and deduce $Q(x)$. But what about the x? We want this to work for *all* x. We accomplish this by fixing x to be an arbitrary element (of the sort we are interested in).

Here are a few examples. First, we will set up the proof structure for a direct proof, then fill in the details.

Example B.5.2

Prove: For all integers n, if n is even, then n^2 is even.

Solution. The format of the proof will be this: Let n be an arbitrary integer. Assume that n is even. Explain explain explain. Therefore n^2 is even.

To fill in the details, we will basically just explain what it means for n to be even, and then see what that means for n^2. Here is a complete proof.

Proof. Let n be an arbitrary integer. Suppose n is even. Then $n = 2k$ for some integer k. Now $n^2 = (2k)^2 = 4k^2 = 2(2k^2)$. Since $2k^2$ is an integer, n^2 is even. ∎

Example B.5.3

Prove: For all integers a, b, and c, if $a|b$ and $b|c$ then $a|c$. (Here $x|y$, read "x divides y" means that y is a multiple of x, i.e., that x will divide into y without remainder).

Solution. Even before we know what the divides symbol means, we can set up a direct proof for this statement. It will go something like this: Let a, b, and c be arbitrary integers. Assume that $a|b$ and $b|c$. Dot dot dot. Therefore $a|c$.

How do we connect the dots? We say what our hypothesis ($a|b$ and $b|c$) really means and why this gives us what the conclusion ($a|c$) really means. Another way to say that $a|b$ is to say that $b = ka$ for some integer k (that is, that b is a multiple of a). What are we going for? That $c = la$, for some integer l (because we want c to be a multiple of a). Here is the complete proof.

Proof. Let a, b, and c be integers. Assume that $a|b$ and $b|c$. In other words, b is a multiple of a and c is a multiple of b. So there are integers k and j such that $b = ka$ and $c = jb$. Combining these (through substitution) we get that $c = jka$. But jk is an integer, so this says that c is a multiple of a. Therefore $a|c$. ∎

B.5.2 Proof by Contrapositive

Recall that an implication $P \to Q$ is logically equivalent to its contrapositive $\neg Q \to \neg P$. There are plenty of examples of statements which are hard to prove directly, but whose contrapositive can easily be proved directly. This is all that **proof by contrapositive** does. It gives a direct proof of the contrapositive of the implication. This is enough because the contrapositive is logically equivalent to the original implication.

The skeleton of the proof of $P \to Q$ by contrapositive will always look roughly like this:

Assume $\neg Q$. Explain, explain, ... explain. Therefore $\neg P$.

As before, if there are variables and quantifiers, we set them to be arbitrary elements of our domain. Here are two examples:

Example B.5.4

Is the statement "for all integers n, if n^2 is even, then n is even" true?

Solution. This is the converse of the statement we proved above using a direct proof. From trying a few examples, this statement definitely appears to be true. So let's prove it.

A direct proof of this statement would require fixing an arbitrary n and assuming that n^2 is even. But it is not at all clear how this would allow us to conclude anything about n. Just because $n^2 = 2k$ does not in itself suggest how we could write n as a multiple of 2.

Try something else: write the contrapositive of the statement. We get, for all integers n, if n is odd then n^2 is odd. This looks much more promising. Our proof will look something like this:

Let n be an arbitrary integer. Suppose that n is not even. This means that.... In other words.... But this is the same as saying.... Therefore n^2 is not even.

Now we fill in the details:

Proof. We will prove the contrapositive. Let n be an arbitrary integer. Suppose that n is not even, and thus odd. Then $n = 2k + 1$ for some integer k. Now $n^2 = (2k+1)^2 = 4k^2 + 4k + 1 = 2(2k^2 + 2k) + 1$. Since $2k^2 + 2k$ is an integer, we see that n^2 is odd and therefore not even. ∎

Example B.5.5
Prove: for all integers a and b, if $a + b$ is odd, then a is odd or b is odd.

Solution. The problem with trying a direct proof is that it will be hard to separate a and b from knowing something about $a + b$. On the other hand, if we know something about a and b separately, then combining them might give us information about $a + b$. The contrapositive of the statement we are trying to prove is: for all integers a and b, if a and b are even, then $a + b$ is even. Thus our proof will have the following format:

Let a and b be integers. Assume that a and b are both even. la la la. Therefore $a + b$ is even.

Here is a complete proof:

Proof. Let a and b be integers. Assume that a and b are even. Then $a = 2k$ and $b = 2l$ for some integers k and l. Now $a + b = 2k + 2l = 2(k + l)$. Since $k + l$ is an integer, we see that $a + b$ is even, completing the proof. ∎

Note that our assumption that a and b are even is really the negation of a or b is odd. We used De Morgan's law here.

We have seen how to prove some statements in the form of implications: either directly or by contrapositive. Some statements are not written as implications to begin with.

Example B.5.6
Consider the following statement: for every prime number p, either $p = 2$ or p is odd. We can rephrase this: for every prime number p, if $p \neq 2$, then p is odd. Now try to prove it.

Solution.
Proof. Let p be an arbitrary prime number. Assume p is not odd. So p is divisible by 2. Since p is prime, it must have exactly two divisors, and it has 2 as a divisor, so p must be divisible by only 1 and 2. Therefore $p = 2$. This completes the proof (by contrapositive). ∎

B.5.3 Proof by Contradiction

There might be statements which really cannot be rephrased as implications. For example, "$\sqrt{2}$ is irrational." In this case, it is hard to know where to start. What can we assume? Well, say we want to prove the statement P. What if we could prove that $\neg P \to Q$ where Q was false? If this implication is true, and Q is false, what can we say about $\neg P$? It must be false as well, which makes P true!

This is why **proof by contradiction** works. If we can prove that $\neg P$ leads to a contradiction, then the only conclusion is that $\neg P$ is false, so P is true. That's what we wanted to prove. In other words, if it is impossible for P to be false, P must be true.

Here are three examples of proofs by contradiction:

Example B.5.7
Prove that $\sqrt{2}$ is irrational.

Solution.
Proof. Suppose not. Then $\sqrt{2}$ is equal to a fraction $\frac{a}{b}$. Without loss of generality, assume $\frac{a}{b}$ is in lowest terms (otherwise reduce the fraction). So,
$$2 = \frac{a^2}{b^2}$$
$$2b^2 = a^2.$$

Thus a^2 is even, and as such a is even. So $a = 2k$ for some integer k, and $a^2 = 4k^2$. We then have,
$$2b^2 = 4k^2$$
$$b^2 = 2k^2.$$

Thus b^2 is even, and as such b is even. Since a is also even, we see that $\frac{a}{b}$ is not in lowest terms, a contradiction. Thus $\sqrt{2}$ is irrational. ∎

Example B.5.8
Prove: There are no integers x and y such that $x^2 = 4y + 2$.

Solution.
Proof. We proceed by contradiction. So suppose there *are* integers x and y such that $x^2 = 4y + 2 = 2(2y + 1)$. So x^2 is even. We have seen that this implies that x is even. So $x = 2k$ for some integer k. Then $x^2 = 4k^2$. This in turn gives $2k^2 = (2y + 1)$. But $2k^2$ is even, and $2y + 1$ is odd, so these cannot be equal. Thus we have a contradiction, so there must not be any integers x and y such that $x^2 = 4y + 2$. ∎

Example B.5.9
The **Pigeonhole Principle**: If more than n pigeons fly into n pigeon holes, then at least one pigeon hole will contain at least two pigeons. Prove this!

Solution.
Proof. Suppose, contrary to stipulation, that each of the pigeon holes contains at most one pigeon. Then at most there will be n pigeons. But we assumed that there are more than n pigeons, so this is impossible. Thus there must be a pigeonhole with more than one pigeon. ∎

> While we phrased this proof as a proof by contradiction, we could have also used a proof by contrapositive since our contradiction was simply the negation of the hypothesis. Sometimes this will happen, in which case you can use either style of proof. There are examples however where the contradiction occurs "far away" from the original statement.

B.5.4 Proof by (counter) Example

It is almost NEVER okay to prove a statement with just an example. Certainly none of the statements proved above can be proved through an example. This is because in each of those cases we are trying to prove that something holds of all integers. We claim that n^2 being even implies that n is even, *no matter what integer n we pick*. Showing that this works for $n = 4$ is not even close to enough.

This cannot be stressed enough. If you are trying to prove a statement of the form $\forall x P(x)$, you absolutely CANNOT prove this with an example.[4]

However, existential statements can be proven this way. If we want to prove that there is an integer n such that $n^2 - n + 41$ is not prime, all we need to do is find one. This might seem like a silly thing to want to prove until you try a few values for n.

n	1	2	3	4	5	6	7
$n^2 - n + 41$	41	43	47	53	61	71	83

So far we have gotten only primes. You might be tempted to conjecture, "For all positive integers n, the number $n^2 - n + 41$ is prime." If you wanted to prove this, you would need to use a direct proof, a proof by contrapositive, or another style of proof, but certainly it is not enough to give even 7 examples. In fact, we can prove this conjecture is *false* by proving its negation: "There is a positive integer n such that $n^2 - n + 41$ is not prime." Since this is an existential statement, it suffices to show that there does indeed exist such a number.

In fact, we can quickly see that $n = 41$ will give 41^2 which is certainly not prime. You might say that this is a counterexample to the conjecture that $n^2 - n + 41$ is always prime. Since so many statements in mathematics are universal, making their negations existential, we can often prove that a statement is false (if it is) by providing a counterexample.

> **Example B.5.10**
>
> Above we proved, "for all integers a and b, if $a + b$ is odd, then a is odd or b is odd." Is the converse true?
>
> **Solution.** The converse is the statement, "for all integers a and b, if a is odd or b is odd, then $a + b$ is odd." This is false! How do we prove it is false? We need to prove the negation of the converse. Let's look at the symbols. The converse is
>
> $$\forall a \forall b ((O(a) \vee O(b)) \rightarrow O(a + b)).$$

---

[4]This is not to say that looking at examples is a waste of time. Doing so will often give you an idea of how to write a proof. But the examples do not belong in the proof.

We want to prove the negation:

$$\neg \forall a \forall b ((O(a) \vee O(b)) \to O(a+b)).$$

Simplify using the rules from the previous sections:

$$\exists a \exists b ((O(a) \vee O(b)) \wedge \neg O(a+b)).$$

As the negation passed by the quantifiers, they changed from $\forall$ to $\exists$. We then needed to take the negation of an implication, which is equivalent to asserting the if part and not the then part.

Now we know what to do. To prove that the converse is false we need to find two integers a and b so that a is odd or b is odd, but $a+b$ is not odd (so even). That's easy: 1 and 3. (remember, "or" means one or the other or both). Both of these are odd, but $1+3=4$ is not odd.

B.5.5 Proof by Cases

We could go on and on and on about different proof styles (we haven't even mentioned induction or combinatorial proofs here), but instead we will end with one final useful technique: proof by cases. The idea is to prove that P is true by proving that $Q \to P$ and $\neg Q \to P$ for some statement Q. So no matter what, whether or not Q is true, we know that P is true. In fact, we could generalize this. Suppose we want to prove P. We know that at least one of the statements $Q_1, Q_2, \ldots, Q_n$ is true. If we can show that $Q_1 \to P$ and $Q_2 \to P$ and so on all the way to $Q_n \to P$, then we can conclude P. The key thing is that we want to be sure that one of our cases (the Q_i's) must be true no matter what.

If that last paragraph was confusing, perhaps an example will make things better.

Example B.5.11

Prove: For any integer n, the number $(n^3 - n)$ is even.

Solution. It is hard to know where to start this, because we don't know much of anything about n. We might be able to prove that $n^3 - n$ is even if we knew that n was even. In fact, we could probably prove that $n^3 - n$ was even if n was odd. But since n must either be even or odd, this will be enough. Here's the proof.

Proof. We consider two cases: if n is even or if n is odd.
 Case 1: n is even. Then $n = 2k$ for some integer k. This give

$$n^3 - n = 8k^3 - 2k$$
$$= 2(4k^2 - k),$$

and since $4k^2 - k$ is an integer, this says that $n^3 - n$ is even.
 Case 2: n is odd. Then $n = 2k + 1$ for some integer k. This gives

$$n^3 - n = (2k+1)^3 - (2k+1)$$
$$= 8k^3 + 6k^2 + 6k + 1 - 2k - 1$$
$$= 2(4k^3 + 3k^2 + 2k),$$

and since $4k^3 + 3k^2 + 2k$ is an integer, we see that $n^3 - n$ is even again.
 Since $n^3 - n$ is even in both exhaustive cases, we see that $n^3 - n$ is indeed always even. ∎

B.5.6 Exercises

1. Consider the statement "for all integers a and b, if $a + b$ is even, then a and b are even"

    (a) Write the contrapositive of the statement.

    (b) Write the converse of the statement.

    (c) Write the negation of the statement.

    (d) Is the original statement true or false? Prove your answer.

    (e) Is the contrapositive of the original statement true or false? Prove your answer.

    (f) Is the converse of the original statement true or false? Prove your answer.

    (g) Is the negation of the original statement true or false? Prove your answer.

    [Solution]

2. For each of the statements below, say what method of proof you should use to prove them. Then say how the proof starts and how it ends. Bonus points for filling in the middle.

    (a) There are no integers x and y such that x is a prime greater than 5 and $x = 6y + 3$.

    (b) For all integers n, if n is a multiple of 3, then n can be written as the sum of consecutive integers.

    (c) For all integers a and b, if $a^2 + b^2$ is odd, then a or b is odd.

    [Solution]

3. Consider the statement: for all integers n, if n is even then $8n$ is even.

   (a) Prove the statement. What sort of proof are you using?

   (b) Is the converse true? Prove or disprove.

   [Solution]

4. The game TENZI comes with 40 six-sided dice (each numbered 1 to 6). Suppose you roll all 40 dice.

   (a) Prove that there will be at least seven dice that land on the same number.

   (b) How many dice would you have to roll before you were guaranteed that some four of them would all match or all be different? Prove your answer.

   [Solution]

5. Prove that for all integers n, it is the case that n is even if and only if $3n$ is even. That is, prove both implications: if n is even, then $3n$ is even, and if $3n$ is even, then n is even.

   Hint. One of the implications will be a direct proof, the other will be a proof by contrapositive.

6. Prove that $\sqrt{3}$ is irrational.

   Hint. This is really an exercise in modifying the proof that $\sqrt{2}$ is irrational. There you proved things were even; here they will be multiples of 3.

7. Consider the statement: for all integers a and b, if a is even and b is a multiple of 3, then ab is a multiple of 6.

   (a) Prove the statement. What sort of proof are you using?

   (b) State the converse. Is it true? Prove or disprove.

   Hint. Part (a) should be a relatively easy direct proof. Look for a counterexample for part (b).

8. Prove the statement: For all integers n, if $5n$ is odd, then n is odd. Clearly state the style of proof you are using.

9. Prove the statement: For all integers a, b, and c, if $a^2 + b^2 = c^2$, then a or b is even.

   Hint. A proof by contradiction would be reasonable here, because then you get to assume that both a and b are odd. Deduce that c^2 is even, and therefore a multiple of 4 (why? and why is that a contradiction?).

10. Suppose that you would like to prove the following implication:

    For all numbers n, if n is prime then n is solitary.

    Write out the beginning and end of the argument if you were to prove the statement,

    (a) Directly

    (b) By contrapositive

    (c) By contradiction

You do not need to provide details for the proofs (since you do not know what solitary means). However, make sure that you provide the first few and last few lines of the proofs so that we can see that logical structure you would follow.

11. Suppose you have a collection of 5-cent stamps and 8-cent stamps. We saw earlier that it is possible to make any amount of postage greater than 27 cents using combinations of both these types of stamps. But, let's ask some other questions:

    (a) Prove that if you only use an even number of both types of stamps, the amount of postage you make must be even.

    (b) Suppose you made an even amount of postage. Prove that you used an even number of at least one of the types of stamps.

    (c) Suppose you made exactly 72 cents of postage. Prove that you used at least 6 of one type of stamp.

    Hint. Use a different style of proof for each part. The last part should remind you of the pigeonhole principle, so mimicking that proof might be helpful.

12. Prove: $x = y$ if and only if $xy = \dfrac{(x+y)^2}{4}$. Note, you will need to prove two "directions" here: the "if" and the "only if" part.

13. Prove that $\log(7)$ is irrational.

    Hint. Note that if $\log(7) = \frac{a}{b}$, then $7 = 10^{\frac{a}{b}}$. Can any power of 7 be the same as a power of 10?

14. Prove that there are no integer solutions to the equation $x^2 = 4y + 3$.

    Hint. What if there were? Deduce that x must be odd, and continue towards a contradiction.

15. Prove that every prime number greater than 3 is either one more or one less than a multiple of 6.

    Hint. Prove the contrapositive by cases. There will be 4 cases to consider.

16. Your "friend" has shown you a "proof" he wrote to show that $1 = 3$. Here is the proof:

    Proof. I claim that $1 = 3$. Of course we can do anything to one side of an equation as long as we also do it to the other side. So subtract 2 from both sides. This gives $-1 = 1$. Now square both sides, to get $1 = 1$. And we all agree this is true. ■

    What is going on here? Is your friend's argument valid? Is the argument a proof of the claim $1 = 3$? Carefully explain using what we know about logic.

    Hint. Your friend's proof a proof, but of what? What implication follows from the given proof? Is that helpful?

17. A standard deck of 52 cards consists of 4 suites (hearts, diamonds, spades and clubs) each containing 13 different values (Ace, 2, 3, ..., 10, J, Q, K). If

you draw some number of cards at random you might or might not have a pair (two cards with the same value) or three cards all of the same suit. However, if you draw enough cards, you will be guaranteed to have these. For each of the following, find the smallest number of cards you would need to draw to be guaranteed having the specified cards. Prove your answers.

(a) Three of a kind (for example, three 7's).

(b) A flush of five cards (for example, five hearts).

(c) Three cards that are either all the same suit or all different suits.

18. Suppose you are at a party with 19 of your closest friends (so including you, there are 20 people there). Explain why there must be least two people at the party who are friends with the same number of people at the party. Assume friendship is always reciprocated.

Hint. Consider the set of *numbers* of friends that everyone has. If everyone had a different number of friends, this set must contain 20 elements. Is that possible? Why not?

19. Your friend has given you his list of 115 best Doctor Who episodes (in order of greatness). It turns out that you have seen 60 of them. Prove that there are at least two episodes you have seen that are exactly four episodes apart on your friend's list.

Hint. This feels like the pigeonhole principle, although a bit more complicated. At least, you could try to replicate the style of proof used by the pigeonhole principle. How would the episodes need to be spaced out so that no two of your sixty were exactly 4 apart?

20. Suppose you have an $n \times n$ chessboard but your dog has eaten one of the corner squares. Can you still cover the remaining squares with dominoes? What needs to be true about n? Give necessary and sufficient conditions (that is, say exactly which values of n work and which do not work). Prove your answers.

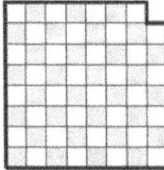

21. What if your $n \times n$ chessboard is missing two opposite corners? Prove that no matter what n is, you will not be able to cover the remaining squares with dominoes.

B.6 Induction

Mathematical induction is a proof technique, not unlike direct proof or proof by contradiction or combinatorial proof. In other words, induction is a style of argument we use to convince ourselves and others that a mathematical statement is always true. Many mathematical statements can be proved by simply explaining what they mean. Others are very difficult to prove—in fact, there are relatively simple mathematical statements which nobody yet knows how to prove. To facilitate the discovery of proofs, it is important to be familiar with some standard styles of arguments. Induction is one such style. Let's start with an example:

B.6.1 Stamps

> **Puzzle 337**
> You need to mail a package, but don't yet know how much postage you will need. You have a large supply of 8-cent stamps and 5-cent stamps. Which amounts of postage can you make exactly using these stamps? Which amounts are impossible to make?

Perhaps in investigating the problem above you picked some amounts of postage, and then figured out whether you could make that amount using just 8-cent and 5-cent stamps. Perhaps you did this in order: can you make 1 cent of postage? Can you make 2 cents? 3 cents? And so on. If this is what you did, you were actually answering a *sequence* of questions. We have methods for dealing with sequences. Let's see if that helps.

Actually, we will not make a sequence of questions, but rather a sequence of statements. Let $P(n)$ be the statement "you can make n cents of postage using just 8-cent and 5-cent stamps." Since for each value of n, $P(n)$ is a statement, it is either true or false. So if we form the sequence of statements

$$P(1), P(2), P(3), P(4), \ldots,$$

the sequence will consist of T's (for true) and F's (for false). In our particular case the sequence starts

$$F, F, F, F, T, F, F, T, F, T, F, F, T, \ldots$$

because $P(1), P(2), P(3), P(4)$ are all false (you cannot make 1, 2, 3, or 4 cents of postage) but $P(5)$ is true (use one 5-cent stamp), and so on.

Let's think a bit about how we could find the value of $P(n)$ for some specific n (the "value" will be either T or F). How did we find the value of the nth term of a sequence of numbers? How did we find a_n? There were two ways we could do this: either there was a closed formula for a_n, so we could plug in n into the formula and get our output value, or we had a recursive definition for the sequence, so we could use the previous terms of the sequence to compute the nth term. When dealing with sequences of statements, we could use either of these techniques as well. Maybe there is a way to use n itself to determine whether we can make n cents of postage. That would be something like a closed formula. Or instead we could use the previous terms in the sequence (of statements) to determine whether we can make n cents of postage. That is, if we know the value of $P(n-1)$, can we get from that to the value of $P(n)$? That would be something like a recursive definition for the sequence. Remember, finding recursive definitions for sequences was often easier than finding closed formulas. The same is true here.

Suppose I told you that $P(43)$ was true (it is). Can you determine from this fact the value of $P(44)$ (whether it true or false)? Yes you can. Even if we don't know how exactly we made 43 cents out of the 5-cent and 8-cent stamps, we do know that there was some way to do it. What if that way used at least three 5-cent stamps (making 15 cents)? We could replace those three 5-cent stamps with two 8-cent stamps (making 16 cents). The total postage has gone up by 1, so we have a way to make 44 cents, so $P(44)$ is true. Of course, we assumed that we had at least three 5-cent stamps. What if we didn't? Then we must have at least three 8-cent stamps (making 24 cents). If we replace those three 8-cent stamps with five 5-cent stamps (making 25 cents) then again we have bumped up our total by 1 cent so we can make 44 cents, so $P(44)$ is true.

Notice that we have not said how to make 44 cents, just that we can, on the basis that we can make 43 cents. How do we know we can make 43 cents? Perhaps because we know we can make 42 cents, which we know we can do because we know we can make 41 cents, and so on. It's a recursion! As with a recursive definition of a numerical sequence, we must specify our initial value. In this case, the initial value is "$P(1)$ is false." That's not good, since our recurrence relation just says that $P(k+1)$ is true *if* $P(k)$ is also true. We need to start the process with a true $P(k)$. So instead, we might want to use "$P(28)$ is true" as the initial condition.

Putting this all together we arrive at the following fact: it is possible to (exactly) make any amount of postage greater than 27 cents using just 5-cent and 8-cent stamps.[5] In other words, $P(k)$ is true for any $k \geq 28$. To prove this, we could do the following:

1. Demonstrate that $P(28)$ is true.

2. Prove that if $P(k)$ is true, then $P(k+1)$ is true (for any $k \geq 28$).

Suppose we have done this. Then we know that the 28th term of the sequence above is a T (using step 1, the initial condition or **base case**), and that every term after the 28th is T also (using step 2, the recursive part or **inductive case**). Here is what the proof would actually look like.

Proof. Let $P(n)$ be the statement "it is possible to make exactly n cents of postage using 5-cent and 8-cent stamps." We will show $P(n)$ is true for all $n \geq 28$.

First, we show that $P(28)$ is true: $28 = 4 \cdot 5 + 1 \cdot 8$, so we can make 28 cents using four 5-cent stamps and one 8-cent stamp.

Now suppose $P(k)$ is true for some arbitrary $k \geq 28$. Then it is possible to make k cents using 5-cent and 8-cent stamps. Note that since $k \geq 28$, it cannot be that we use fewer than three 5-cent stamps *and* fewer than three 8-cent stamps: using two of each would give only 26 cents. Now if we have made k cents using at least three 5-cent stamps, replace three 5-cent stamps by two 8-cent stamps. This replaces 15 cents of postage with 16 cents, moving from a total of k cents to

---

[5]This is not claiming that there are no amounts less than 27 cents which can also be made.

$k + 1$ cents. Thus $P(k + 1)$ is true. On the other hand, if we have made k cents using at least three 8-cent stamps, then we can replace three 8-cent stamps with five 5-cent stamps, moving from 24 cents to 25 cents, giving a total of $k + 1$ cents of postage. So in this case as well $P(k + 1)$ is true.

Therefore, by the principle of mathematical induction, $P(n)$ is true for all $n \geq 28$. ∎

B.6.2 Formalizing Proofs

What we did in the stamp example above works for many types of problems. Proof by induction is useful when trying to prove statements about all natural numbers, or all natural numbers greater than some fixed first case (like 28 in the example above), and in some other situations too. In particular, induction should be used when there is some way to go from one case to the next – when you can see how to always "do one more."

This is a big idea. Thinking about a problem *inductively* can give new insight into the problem. For example, to really understand the stamp problem, you should think about how any amount of postage (greater than 28 cents) can be made (this is non-inductive reasoning) and also how the ways in which postage can be made *changes* as the amount increases (inductive reasoning). When you are asked to provide a proof by induction, you are being asked to think about the problem *dynamically*; how does increasing n change the problem?

But there is another side to proofs by induction as well. In mathematics, it is not enough to understand a problem, you must also be able to communicate the problem to others. Like any discipline, mathematics has standard language and style, allowing mathematicians to share their ideas efficiently. Proofs by induction have a certain formal style, and being able to write in this style is important. It allows us to keep our ideas organized and might even help us with formulating a proof.

Here is the general structure of a proof by mathematical induction:

Induction Proof Structure.

Start by saying what the statement is that you want to prove: "Let $P(n)$ be the statement..." To prove that $P(n)$ is true for all $n \geq 0$, you must prove two facts:

1. Base case: Prove that $P(0)$ is true. You do this directly. This is often easy.

2. Inductive case: Prove that $P(k) \to P(k + 1)$ for all $k \geq 0$. That is, prove that for any $k \geq 0$ if $P(k)$ is true, then $P(k + 1)$ is true as well. This is the proof of an if... then... statement, so you can assume $P(k)$ is true ($P(k)$ is called the *inductive hypothesis*). You must then explain why $P(k + 1)$ is also true, given that assumption.

Assuming you are successful on both parts above, you can conclude, "Therefore by the principle of mathematical induction, the statement $P(n)$ is true for all $n \geq 0$."

Sometimes the statement $P(n)$ will only be true for values of $n \geq 4$, for example, or some other value. In such cases, replace all the 0's above with 4's (or the other value).

The other advantage of formalizing inductive proofs is it allows us to verify that the logic behind this style of argument is valid. Why does induction work? Think of a row of dominoes set up standing on their edges. We want to argue that in a minute, all the dominoes will have fallen down. For this to happen, you will need to push the first domino. That is the base case. It will also have to be that the dominoes are close enough together that when any particular domino falls, it will cause the next domino to fall. That is the inductive case. If both of these conditions are met, you push the first domino over and each domino will cause the next to fall, then all the dominoes will fall.

Induction is powerful! Think how much easier it is to knock over dominoes when you don't have to push over each domino yourself. You just start the chain reaction, and the rely on the relative nearness of the dominoes to take care of the rest.

Think about our study of sequences. It is easier to find recursive definitions for sequences than closed formulas. Going from one case to the next is easier than going directly to a particular case. That is what is so great about induction. Instead of going directly to the (arbitrary) case for n, we just need to say how to get from one case to the next.

When you are asked to prove a statement by mathematical induction, you should first think about *why* the statement is true, using inductive reasoning. Explain why induction is the right thing to do, and roughly why the inductive case will work. Then, sit down and write out a careful, formal proof using the structure above.

B.6.3 Examples

Here are some examples of proof by mathematical induction.

> **Example B.6.1**
> Prove for each natural number $n \geq 1$ that $1 + 2 + 3 + \cdots + n = \frac{n(n+1)}{2}$.
>
> **Solution.** First, let's think inductively about this equation. In fact, we know this is true for other reasons (reverse and add comes to mind). But why might induction be applicable? The left-hand side adds up the numbers from 1 to n. If we know how to do that, adding just one more term $(n + 1)$ would not be that hard. For example, if $n = 100$, suppose we know that the sum of the first 100 numbers is 5050 (so $1+2+3+\cdots+100 = 5050$, which is true). Now to find the sum of the first 101 numbers, it makes more sense to just add 101 to 5050, instead of computing the entire sum again. We would have $1 + 2 + 3 + \cdots + 100 + 101 = 5050 + 101 = 5151$. In fact, it would always be easy to add just one more term. This is why we should use induction.
>
> Now the formal proof:

Proof. Let $P(n)$ be the statement $1 + 2 + 3 + \cdots + n = \frac{n(n+1)}{2}$. We will show that $P(n)$ is true for all natural numbers $n \geq 1$.

Base case: $P(1)$ is the statement $1 = \frac{1(1+1)}{2}$ which is clearly true.

Inductive case: Let $k \geq 1$ be a natural number. Assume (for induction) that $P(k)$ is true. That means $1 + 2 + 3 + \cdots + k = \frac{k(k+1)}{2}$. We will prove that $P(k+1)$ is true as well. That is, we must prove that $1 + 2 + 3 + \cdots + k + (k+1) = \frac{(k+1)(k+2)}{2}$. To prove this equation, start by adding $k+1$ to both sides of the inductive hypothesis:

$$1 + 2 + 3 + \cdots + k + (k+1) = \frac{k(k+1)}{2} + (k+1).$$

Now, simplifying the right side we get:

$$\frac{k(k+1)}{2} + k + 1 = \frac{k(k+1)}{2} + \frac{2(k+1)}{2}$$
$$= \frac{k(k+1) + 2(k+1)}{2}$$
$$= \frac{(k+2)(k+1)}{2}.$$

Thus $P(k+1)$ is true, so by the principle of mathematical induction $P(n)$ is true for all natural numbers $n \geq 1$. ∎

Note that in the part of the proof in which we proved $P(k+1)$ from $P(k)$, we used the equation $P(k)$. This was the inductive hypothesis. Seeing how to use the inductive hypotheses is usually straight forward when proving a fact about a sum like this. In other proofs, it can be less obvious where it fits in.

Example B.6.2

Prove that for all $n \in \mathbb{N}$, $6^n - 1$ is a multiple of 5.

Solution. Again, start by understanding the dynamics of the problem. What does increasing n do? Let's try with a few examples. If $n = 1$, then yes, $6^1 - 1 = 5$ is a multiple of 5. What does incrementing n to 2 look like? We get $6^2 - 1 = 35$, which again is a multiple of 5. Next, $n = 3$: but instead of just finding $6^3 - 1$, what did the increase in n do? We will still subtract 1, but now we are multiplying by another 6 first. Viewed another way, we are multiplying a number which is one more than a multiple of 5 by 6 (because $6^2 - 1$ is a multiple of 5, so 6^2 is one more than a multiple of 5). What do numbers which are one more than a multiple of 5 look like? They must have last digit 1 or 6. What happens when you multiply such a number by 6? Depends on the number, but in any case, the last digit of the new number must be a 6. And then if you subtract 1, you get last digit 5, so a multiple of 5.

The point is, every time we multiply by just one more six, we still get a number with last digit 6, so subtracting 1 gives us a multiple of 5. Now the formal proof:

Proof. Let $P(n)$ be the statement, "$6^n - 1$ is a multiple of 5." We will prove that $P(n)$ is true for all $n \in \mathbb{N}$.

Base case: $P(0)$ is true: $6^0 - 1 = 0$ which is a multiple of 5.

Inductive case: Let k be an arbitrary natural number. Assume, for induction, that $P(k)$ is true. That is, $6^k - 1$ is a multiple of 5. Then $6^k - 1 = 5j$ for some integer j. This means that $6^k = 5j + 1$. Multiply both sides by 6:

$$6^{k+1} = 6(5j + 1) = 30j + 6.$$

But we want to know about $6^{k+1} - 1$, so subtract 1 from both sides:

$$6^{k+1} - 1 = 30j + 5.$$

Of course $30j + 5 = 5(6j + 1)$, so is a multiple of 5.

Therefore $6^{k+1} - 1$ is a multiple of 5, or in other words, $P(k + 1)$ is true. Thus, by the principle of mathematical induction $P(n)$ is true for all $n \in \mathbb{N}$. ∎

We had to be a little bit clever (i.e., use some algebra) to locate the $6^k - 1$ inside of $6^{k+1} - 1$ before we could apply the inductive hypothesis. This is what can make inductive proofs challenging.

In the two examples above, we started with $n = 1$ or $n = 0$. We can start later if we need to.

Example B.6.3

Prove that $n^2 < 2^n$ for all integers $n \geq 5$.

Solution. First, the idea of the argument. What happens when we increase n by 1? On the left-hand side, we increase the base of the square and go to the next square number. On the right-hand side, we increase the power of 2. This means we double the number. So the question is, how does doubling a number relate to increasing to the next square? Think about what the difference of two consecutive squares looks like. We have $(n + 1)^2 - n^2$. This factors:

$$(n + 1)^2 - n^2 = (n + 1 - n)(n + 1 + n) = 2n + 1.$$

But doubling the right-hand side increases it by 2^n, since $2^{n+1} = 2^n + 2^n$. When n is large enough, $2^n > 2n + 1$.

What we are saying here is that each time n increases, the left-hand side grows by less than the right-hand side. So if the left-hand side starts smaller (as it does when $n = 5$), it will never catch up. Now the formal proof:

Proof. Let $P(n)$ be the statement $n^2 < 2^n$. We will prove $P(n)$ is true for all integers $n \geq 5$.

Base case: $P(5)$ is the statement $5^2 < 2^5$. Since $5^2 = 25$ and $2^5 = 32$, we see that $P(5)$ is indeed true.

Inductive case: Let $k \geq 5$ be an arbitrary integer. Assume, for induction, that $P(k)$ is true. That is, assume $k^2 < 2^k$. We will prove that $P(k+1)$ is true, i.e., $(k+1)^2 < 2^{k+1}$. To prove such an inequality, start with the left-hand side and work towards the right-hand side:

$$\begin{aligned}
(k+1)^2 &= k^2 + 2k + 1 \\
&< 2^k + 2k + 1 \quad \ldots \text{by the inductive hypothesis.} \\
&< 2^k + 2^k \quad \ldots \text{since } 2k + 1 < 2^k \text{ for } k \geq 5. \\
&= 2^{k+1}.
\end{aligned}$$

Following the equalities and inequalities through, we get $(k+1)^2 < 2^{k+1}$, in other words, $P(k+1)$. Therefore by the principle of mathematical induction, $P(n)$ is true for all $n \geq 5$. ∎

The previous example might remind you of the *racetrack principle* from calculus, which says that if $f(a) < g(a)$, and $f'(x) < g'(x)$ for $x > a$, then $f(x) < g(x)$ for $x > a$. Same idea: the larger function is increasing at a faster rate than the smaller function, so the larger function will stay larger. In discrete math, we don't have derivatives, so we look at differences. Thus induction is the way to go.

WARNING:.

With great power, comes great responsibility. Induction isn't magic. It seems very powerful to be able to assume $P(k)$ is true. After all, we are trying to prove $P(n)$ is true and the only difference is in the variable: k vs. n. Are we assuming that what we want to prove is true? Not really. We assume $P(k)$ is true only for the sake of proving that $P(k+1)$ is true.

Still you might start to believe that you can prove anything with induction. Consider this incorrect "proof" that every Canadian has the same eye color: Let $P(n)$ be the statement that any n Canadians have the same eye color. $P(1)$ is true, since everyone has the same eye color as themselves. Now assume $P(k)$ is true. That is, assume that in any group of k Canadians, everyone has the same eye color. Now consider an arbitrary group of $k+1$ Canadians. The first k of these must all have the same eye color, since $P(k)$ is true. Also, the last k of these must have the same eye color, since $P(k)$ is true. So in fact, everyone the group must have the same eye color. Thus $P(k+1)$ is true. So by the principle of mathematical induction, $P(n)$ is true for all n.

Clearly something went wrong. The problem is that the proof that $P(k)$ implies $P(k + 1)$ assumes that $k \geq 2$. We have only shown $P(1)$ is true. In fact, $P(2)$ is false.

B.6.4 Strong Induction

Puzzle 338
Start with a square piece of paper. You want to cut this square into smaller squares, leaving no waste (every piece of paper you end up with must be a square). Obviously it is possible to cut the square into 4 squares. You can also cut it into 9 squares. It turns out you can cut the square into 7 squares (although not all the same size). What other numbers of squares could you end up with?

Sometimes, to prove that $P(k + 1)$ is true, it would be helpful to know that $P(k)$ and $P(k - 1)$ and $P(k - 2)$ are all true. Consider the following puzzle:

> You have a rectangular chocolate bar, made up of n identical squares of chocolate. You can take such a bar and break it along any row or column. How many times will you have to break the bar to reduce it to n single chocolate squares?

At first, this question might seem impossible. Perhaps I meant to ask for the *smallest* number of breaks needed? Let's investigate.

Start with some small cases. If $n = 2$, you must have a 1×2 rectangle, which can be reduced to single pieces in one break. With $n = 3$, we must have a 1×3 bar, which requires two breaks: the first break creates a single square and a 1×2 bar, which we know takes one (more) break.

What about $n = 4$? Now we could have a 2×2 bar, or a 1×4 bar. In the first case, break the bar into two 2×2 bars, each which require one more break (that's a total of three breaks required). If we started with a 1×4 bar, we have choices for our first break. We could break the bar in half, creating two 1×2 bars, or we could break off a single square, leaving a 1×3 bar. But either way, we still need two more breaks, giving a total of three.

It is starting to look like no matter how we break the bar (and no matter how the n squares are arranged into a rectangle), we will always have the same number of breaks required. It also looks like that number is one less than n:

Conjecture B.6.4 *Given a n-square rectangular chocolate bar, it always takes $n - 1$ breaks to reduce the bar to single squares.*

It makes sense to prove this by induction because after breaking the bar once, you are left with *smaller* chocolate bars. Reducing to smaller cases is what induction is all about. We can inductively assume we already know how to deal with these smaller bars. The problem is, if we are trying to prove the inductive case about a $(k + 1)$-square bar, we don't know that after the first break the remaining bar will have k squares. So we really need to assume that our conjecture is true for all cases less than $k + 1$.

Is it valid to make this stronger assumption? Remember, in induction we are attempting to prove that $P(n)$ is true for all n. What if that were not the case?

Then there would be some first n_0 for which $P(n_0)$ was false. Since n_0 is the *first* counterexample, we know that $P(n)$ is true for all $n < n_0$. Now we proceed to prove that $P(n_0)$ is actually true, based on the assumption that $P(n)$ is true for all smaller n.

This is quite an advantage: we now have a stronger inductive hypothesis. We can assume that $P(1), P(2), P(3), \ldots P(k)$ is true, just to show that $P(k+1)$ is true. Previously, we just assumed $P(k)$ for this purpose.

It is slightly easier if we change our variables for strong induction. Here is what the formal proof would look like:

Strong Induction Proof Structure.

Again, start by saying what you want to prove: "Let $P(n)$ be the statement..." Then establish two facts:

1. Base case: Prove that $P(0)$ is true.

2. Inductive case: Assume $P(k)$ is true for all $k < n$. Prove that $P(n)$ is true.

Conclude, "therefore, by strong induction, $P(n)$ is true for all $n > 0$."

Of course, it is acceptable to replace 0 with a larger base case if needed.[6]

Let's prove our conjecture about the chocolate bar puzzle:

Proof. Let $P(n)$ be the statement, "it takes $n-1$ breaks to reduce a n-square chocolate bar to single squares."

Base case: Consider $P(2)$. The squares must be arranged into a 1×2 rectangle, and we require $2 - 1 = 1$ breaks to reduce this to single squares.

Inductive case: Fix an arbitrary $n \geq 2$ and assume $P(k)$ is true for all $k < n$. Consider a n-square rectangular chocolate bar. Break the bar once along any row or column. This results in two chocolate bars, say of sizes a and b. That is, we have an a-square rectangular chocolate bar, a b-square rectangular chocolate bar, and $a + b = n$.

We also know that $a < n$ and $b < n$, so by our inductive hypothesis, $P(a)$ and $P(b)$ are true. To reduce the a-sqaure bar to single squares takes $a - 1$ breaks; to reduce the b-square bar to single squares takes $b - 1$ breaks. Doing this results in our original bar being reduced to single squares. All together it took the initial break, plus the $a - 1$ and $b - 1$ breaks, for a total of $1 + a - 1 + b - 1 = a + b - 1 = n - 1$ breaks. Thus $P(n)$ is true.

Therefore, by strong induction, $P(n)$ is true for all $n \geq 2$. ∎

Here is a more mathematically relevant example:

Example B.6.5

Prove that any natural number greater than 1 is either prime or can be written as the product of primes.

---

[6]Technically, strong induction does not require you to prove a separate base case. This is because when proving the inductive case, you must show that $P(0)$ is true, assuming $P(k)$ is true for all $k < 0$. But this is not any help so you end up proving $P(0)$ anyway. To be on the safe side, we will always include the base case separately.

> **Solution.** First, the idea: if we take some number n, maybe it is prime. If so, we are done. If not, then it is composite, so it is the product of two smaller numbers. Each of these factors is smaller than n (but at least 2), so we can repeat the argument with these numbers. We have reduced to a smaller case.
>
> Now the formal proof:
>
> *Proof.* Let $P(n)$ be the statement, "n is either prime or can be written as the product of primes." We will prove $P(n)$ is true for all $n \geq 2$.
>
> Base case: $P(2)$ is true because 2 is indeed prime.
>
> Inductive case: assume $P(k)$ is true for all $k < n$. We want to show that $P(n)$ is true. That is, we want to show that n is either prime or is the product of primes. If n is prime, we are done. If not, then n has more than 2 divisors, so we can write $n = m_1 \cdot m_2$, with m_1 and m_2 less than n (and greater than 1). By the inductive hypothesis, m_1 and m_2 are each either prime or can be written as the product of primes. In either case, we have that n is written as the product of primes.
>
> Thus by the strong induction, $P(n)$ is true for all $n \geq 2$. ∎

Whether you use regular induction or strong induction depends on the statement you want to prove. If you wanted to be safe, you could always use strong induction. It really is *stronger*, so can accomplish everything "weak" induction can. That said, using regular induction is often easier since there is only one place you can use the induction hypothesis. There is also something to be said for *elegance* in proofs. If you can prove a statement using simpler tools, it is nice to do so.

As a final contrast between the two forms of induction, consider once more the stamp problem. Regular induction worked by showing how to increase postage by one cent (either replacing three 5-cent stamps with two 8-cent stamps, or three 8-cent stamps with five 5-cent stamps). We could give a slightly different proof using strong induction. First, we could show *five* base cases: it is possible to make 28, 29, 30, 31, and 32 cents (we would actually say how each of these is made). Now assume that it is possible to make k cents of postage for all $k < n$ as long as $k \geq 28$. As long as $n > 32$, this means in particular we can make $k = n - 5$ cents. Now add a 5-cent stamp to get make n cents.

B.6.5 Exercises

1. On the way to the market, you exchange your cow for some magic dark chocolate espresso beans. These beans have the property that every night at midnight, each bean splits into two, effectively doubling your collection. You decide to take advantage of this and each morning (around 8am) you eat 5 beans.

    (a) Explain why it is true that *if* at noon on day n you have a number of beans ending in a 5, then at noon on day $n + 1$ you will still have a number of beans ending in a 5.

(b) Why is the previous fact not enough to conclude that you will always have a number of beans ending in a 5? What additional fact would you need?

(c) Assuming you have the additional fact in part (b), and have successfully proved the fact in part (a), how do you know that you will always have a number of beans ending in a 5? Illustrate what is going on by carefully explaining how the two facts above prove that you will have a number of beans ending in a 5 on *day 4* specifically. In other words, explain why induction works in this context.

[Solution]

2. Use induction to prove for all $n \in \mathbb{N}$ that $\sum_{k=0}^{n} 2^k = 2^{n+1} - 1$. [Solution]

3. Prove that $7^n - 1$ is a multiple of 6 for all $n \in \mathbb{N}$. [Solution]

4. Prove that $1 + 3 + 5 + \cdots + (2n - 1) = n^2$ for all $n \geq 1$. [Solution]

5. Prove that $F_0 + F_2 + F_4 + \cdots + F_{2n} = F_{2n+1} - 1$ where F_n is the nth Fibonacci number. [Solution]

6. Prove that $2^n < n!$ for all $n \geq 4$. (Recall, $n! = 1 \cdot 2 \cdot 3 \cdots n$.) [Solution]

7. Prove, by mathematical induction, that $F_0 + F_1 + F_2 + \cdots + F_n = F_{n+2} - 1$, where F_n is the nth Fibonacci number ($F_0 = 0, F_1 = 1$ and $F_n = F_{n-1} + F_{n-2}$).

8. Zombie Euler and Zombie Cauchy, two famous zombie mathematicians, have just signed up for Twitter accounts. After one day, Zombie Cauchy has more followers than Zombie Euler. Each day after that, the number of new followers of Zombie Cauchy is exactly the same as the number of new followers of Zombie Euler (and neither lose any followers). Explain how a proof by mathematical induction can show that on every day after the first day, Zombie Cauchy will have more followers than Zombie Euler. That is, explain what the base case and inductive case are, and why they together prove that Zombie Cauchy will have more followers on the 4th day.

9. Find the largest number of points which a football team cannot get exactly using just 3-point field goals and 7-point touchdowns (ignore the possibilities of safeties, missed extra points, and two point conversions). Prove your answer is correct by mathematical induction.

    Hint. It is not possible to score exactly 11 points. Can you prove that you can score n points for any $n \geq 12$?

10. Prove that the sum of n squares can be found as follows

$$1^2 + 2^2 + 3^2 + \ldots + n^2 = \frac{n(n+1)(2n+1)}{6}.$$

11. Prove that the sum of the interior angles of a convex n-gon is $(n - 2) \cdot 180°$. (A convex n-gon is a polygon with n sides for which each interior angle is less than $180°$.)

    Hint. Start with $(k + 1)$-gon and divide it up into a k-gon and a triangle.

12. What is wrong with the following "proof" of the "fact" that $n + 3 = n + 7$ for all values of n (besides of course that the thing it is claiming to prove is false)?

    Proof. Let $P(n)$ be the statement that $n + 3 = n + 7$. We will prove that $P(n)$ is true for all $n \in \mathbb{N}$. Assume, for induction that $P(k)$ is true. That is, $k + 3 = k + 7$. We must show that $P(k + 1)$ is true. Now since $k + 3 = k + 7$, add 1 to both sides. This gives $k + 3 + 1 = k + 7 + 1$. Regrouping $(k + 1) + 3 = (k + 1) + 7$. But this is simply $P(k + 1)$. Thus by the principle of mathematical induction $P(n)$ is true for all $n \in \mathbb{N}$. ∎

    [Solution]

13. The proof in the previous problem does not work. But if we modify the "fact," we can get a working proof. Prove that $n + 3 < n + 7$ for all values of $n \in \mathbb{N}$. You can do this proof with algebra (without induction), but the goal of this exercise is to write out a valid induction proof. [Solution]

14. Find the flaw in the following "proof" of the "fact" that $n < 100$ for every $n \in \mathbb{N}$.

    Proof. Let $P(n)$ be the statement $n < 100$. We will prove $P(n)$ is true for all $n \in \mathbb{N}$. First we establish the base case: when $n = 0$, $P(n)$ is true, because $0 < 100$. Now for the inductive step, assume $P(k)$ is true. That is, $k < 100$. Now if $k < 100$, then k is some number, like 80. Of course $80 + 1 = 81$ which is still less than 100. So $k + 1 < 100$ as well. But this is what $P(k + 1)$ claims, so we have shown that $P(k) \to P(k + 1)$. Thus by the principle of mathematical induction, $P(n)$ is true for all $n \in \mathbb{N}$. ∎

    [Solution]

15. While the above proof does not work (it better not since the statement it is trying to prove is false!) we can prove something similar. Prove that there is a strictly increasing sequence $a_1, a_2, a_3, \ldots$ of numbers (not necessarily integers) such that $a_n < 100$ for all $n \in \mathbb{N}$. (By **strictly increasing** we mean $a_n < a_{n+1}$ for all n. So each term must be larger than the last.)

    Hint. For the inductive step, you can assume you have a strictly increasing sequence up to a_k where $a_k < 100$. Now you just need to find the next term a_{k+1} so that $a_k < a_{k+1} < 100$. What should a_{k+1} be?

16. What is wrong with the following "proof" of the "fact" that for all $n \in \mathbb{N}$, the number $n^2 + n$ is odd?

    Proof. Let $P(n)$ be the statement "$n^2 + n$ is odd." We will prove that $P(n)$ is true for all $n \in \mathbb{N}$. Suppose for induction that $P(k)$ is true, that is, that $k^2 + k$ is odd. Now consider the statement $P(k + 1)$. Now $(k + 1)^2 + (k + 1) = k^2 + 2k + 1 + k + 1 = k^2 + k + 2k + 2$. By the inductive hypothesis, $k^2 + k$ is odd, and of course $2k + 2$ is even. An odd plus an even is always odd, so therefore $(k+1)^2 + (k+1)$ is odd. Therefore by the principle of mathematical induction, $P(n)$ is true for all $n \in \mathbb{N}$. ∎

    [Solution]

17. Now give a valid proof (by induction, even though you might be able to do so without using induction) of the statement, "for all $n \in \mathbb{N}$, the number $n^2 + n$ is even."

Hint. For the inductive case, you will need to show that $(k+1)^2 + (k+1)$ is even. Factor this out and locate the part of it that is $k^2 + k$. What have you assumed about that quantity?

18. Prove that there is a sequence of positive real numbers $a_0, a_1, a_2, \ldots$ such that the partial sum $a_0 + a_1 + a_2 + \cdots + a_n$ is strictly less than 2 for all $n \in \mathbb{N}$. Hint: think about how you could define what a_{k+1} is to make the induction argument work.

    Hint. This is similar to Exercise B.6.5.15, although there you were showing that a sequence had all its terms less than some value, and here you are showing that the sum is less than som value. But the partial sums forms a sequence, so this is actually very similar.

19. Prove that every positive integer is either a power of 2, or can be written as the sum of distinct powers of 2. [Solution]

20. Prove, using strong induction, that every natural number is either a Fibonacci number or can be written as the *sum* of *distinct* Fibonacci numbers.

    Hint. As with the previous question, we will want to subtract something from n in the inductive step. There we subtracted the largest power of 2 less than n. So what should you subtract here?

    Note, you will still need to take care here that the sum you get from the inductive hypothesis, together with the number you subtracted will be a sum of *distinct* Fibonacci numbers. In fact, you could prove that the Fibonacci numbers in the sum are non-consecutive!

21. Use induction to prove that if n people all shake hands with each other, that the total number of handshakes is $\frac{n(n-1)}{2}$.

    Hint. We have already proved this without using induction, but looking at it inductively sheds light onto the problem (and is fun).

    The question you need to answer to complete the inductive step is, how many new handshakes take place when a person $k+1$ enters the room. Why does adding this give you the correct formula?

22. Suppose that a particular real number x has the property that $x + \frac{1}{x}$ is an integer. Prove that $x^n + \frac{1}{x^n}$ is an integer for all natural numbers n.

    Hint. You will need to use strong induction. For the inductive case, try multiplying $\left(x^k + \frac{1}{x^k}\right)\left(x + \frac{1}{x}\right)$ and collect which terms together are integers.

23. Use induction to prove that $\sum_{k=0}^{n} \binom{n}{k} = 2^n$. That is, the sum of the nth row of Pascal's Triangle is 2^n.

    Hint. Here's the idea: since every entry in Pascal's Triangle is the sum of the two entries above it, we can get the $k+1$st row by adding up all the pairs of entry from the kth row. But doing this uses each entry on the kth row twice. Thus each time we drop to the next row, we double the total. Of course, row 0 has sum $1 = 2^0$ (the base case). Now try to make this precise with a formal induction proof. You will use the fact that $\binom{n}{k} = \binom{n-1}{k-1} + \binom{n-1}{k}$ for the inductive case.

24. Use induction to prove $\binom{4}{0} + \binom{5}{1} + \binom{6}{2} + \cdots + \binom{4+n}{n} = \binom{5+n}{n}$. (This is an example of the hockey stick theorem.)

    Hint. To see why this works, try it on a copy of Pascal's triangle. We are adding up the entries along a diagonal, starting with the 1 on the left-hand side of the 4th row. Suppose we add up the first 5 entries on this diagonal. The claim is that the sum is the entry below and to the left of the last of these 5 entries. Note that if this is true, and we instead add up the first 6 entries, we will need to add the entry one spot to the right of the previous sum. But these two together give the entry below them, which is below and left of the last of the 6 entries on the diagonal. If you follow that, you can see what is going on. But it is not a great proof. A formal induction proof is needed.

25. Use the product rule for logarithms ($\log(ab) = \log(a) + \log(b)$) to prove, by induction on n, that $\log(a^n) = n\log(a)$, for all natural numbers $n \geq 2$. [Solution]

26. Let $f_1, f_2, \ldots, f_n$ be differentiable functions. Prove, using induction, that
    $$(f_1 + f_2 + \cdots + f_n)' = f_1' + f_2' + \cdots + f_n'.$$
    You may assume $(f + g)' = f' + g'$ for any differentiable functions f and g.

    Hint. You are allowed to assume the base case. For the inductive case, group all but the last function together as one sum of functions, then apply the usual sum of derivatives rule, and then the inductive hypothesis.

27. Suppose $f_1, f_2, \ldots, f_n$ are differentiable functions. Use mathematical induction to prove the generalized product rule:
    $$(f_1 f_2 f_3 \cdots f_n)' = f_1' f_2 f_3 \cdots f_n + f_1 f_2' f_3 \cdots f_n + f_1 f_2 f_3' \cdots f_n + \cdots + f_1 f_2 f_3 \cdots f_n'.$$
    You may assume the product rule for two functions is true.

    Hint. For the inductive step, we know by the product rule for two functions that
    $$(f_1 f_2 f_3 \cdots f_k f_{k+1})' = (f_1 f_2 f_3 \cdots f_k)' f_{k+1} + (f_1 f_2 f_3 \cdots f_k) f_{k+1}'.$$
    Then use the inductive hypothesis on the first summand, and distribute.

28. You will prove that the Fibonacci numbers satisfy the identity $F_n^2 + F_{n+1}^2 = F_{2n+1}$. One way to do this is to prove the more general identity,
    $$F_m F_n + F_{m+1} F_{n+1} = F_{m+n+1},$$
    and realize that when $m = n$ we get our desired result.

    Note that we now have two variables, so we want to prove this for all $m \geq 0$ and all $n \geq 0$ at the same time. For each such pair (m, n), let $P(m, n)$ be the statement $F_m F_n + F_{m+1} F_{n+1} = F_{m+n+1}$

    (a) First fix $m = 0$ and give a proof by mathematical induction that $P(0, n)$ holds for all $n \geq 0$. Note this proof will be very easy.

    (b) Now fix an arbitrary n and give a proof by *strong* mathematical induction that $P(m, n)$ holds for all $m \geq 0$.

(c) You can now conclude that $P(m,n)$ holds for all $m, n \geq 0$. Do you believe that? Explain why this sort of induction is valid. For example, why do your proofs above guarantee that $P(2,3)$ is true?

29. Given a square, you can cut the square into smaller squares by cutting along lines parallel to the sides of the original square (these lines do not need to travel the entire side length of the original square). For example, by cutting along the lines below, you will divide a square into 6 smaller squares:

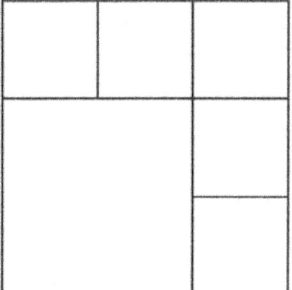

Prove, using strong induction, that it is possible to cut a square into n smaller squares for any $n \geq 6$.

Hint. You will need three base cases. This is a very good hint actually, as it suggests that to prove $P(n)$ is true, you would want to use the fact that $P(n-3)$ is true. So somehow you need to increase the number of squares by 3.

APPENDIX C

HINTS TO SELECTED ACTIVITIES AND EXERCISES

Please resist the urge to look at a hint right away. Try to complete the activity or exercise and only look at the hint for verification or if you are really stuck.

A technical note: if you are using the pdf version, you can click on the title of the hint to link back to the original activity/exercise. Those activities/exercises that have hints will display a link [hint] that will take you to the hint.

Activity 6

(c) This is a little harder. It might help to notice that there is really nothing special about $[n]$ except that it contains n elements. In fact, what can you say about the number of k-element subsets of *any* n-element set? Then, what n-element set would we look at when expanding $(x+y)^n$?

Exercise 9 What can you say about the composition of two bijections?

Activity 11 This should be really easy, but don't let that fool you. Note that a correct proof must start with one side of the identity and manipulate it into the other side. The following is NOT a correct proof, because it starts by assuming the statement you wish to prove.

$$\binom{n}{k} = \binom{n}{n-k}$$
$$\frac{n!}{k!(n-k)!} = \frac{n!}{(n-k)!(n-(n-k)!)}$$
$$\frac{n!}{k!(n-k)!} = \frac{n!}{(n-k)!k!)}$$
$$\frac{n!}{k!(n-k)!} \cdot (k!(n-k)!) = \frac{n!}{(n-k)!(n-(n-k)!)} \cdot (k!(n-k)!)$$
$$n! = n!$$

Activity 12

(a) This is supposed to be very easy.

(b) What if you started with all bow ties in the suitcase and got rid of some? How many do you need to remove?

Activity 13

(d) Recall the bijection principle tells us that if $f : X \to Y$ is a bijection, then $|X| = |Y|$. What is the cardinality of each of the subsets of $\mathcal{P}([n])$ that we are considering?

Activity 14 We say "should", but please don't waste your time doing this all the way out. Unless you are really bored.

Activity 15 Again, use the Pascal recurrence $\binom{n}{k} = \binom{n-1}{k-1} + \binom{n-1}{k}$. Doing this for all summands will give you some repeats.

Activity 16

(a) Your final answer here should be a sum, since we have counted the number of subsets partitioned into disjoint collections. You will be using the sum principle, which we will formally introduce in the next subsection.

(b) Think about how many choices you have for each element. You could put 1 in the subset or not. You could include 2 in the subset or not.

Activity 17 Just do it.

Exercise 23

(a) Many students want this sort of question to be an example of the sum principle (presumably because to get an acceptable path you travel to $(4, 7)$ and then add on a path that gets you the rest of the way. But notice that you are really concatenating two paths, and there are lots of choices for the first path and lots for the second. This sounds more like an ordered pair.

(b) If you knew the answer to this part and the previous part, what is another way you could compute the number of paths from $(0, 0)$ to $(10, 10)$?

(d) Compare to the question about how many playing cards are either red or a face card, and why the answer is not $26 + 12$.

Exercise 24 Try applying the product principle in the case $n = 2$ and $n = 3$. How might you apply it in general?

Exercise 25

(a) Suppose you have a list in alphabetical order of names of the members of the club. In how many ways can you pair up the first person on the list? In how many ways can you pair up the next person who isn't already paired up?

Exercise 26 In how many ways may you assign the men to their rows? The women? Once a woman and a man have a row to share, in how many ways may they choose their seats?

Activity 27

(a) We are not necessarily looking for a numerical answer here, but you should be able to express the answer using notation we have developed already. Or by using some sort of triangle.

(c) Should your answer be larger or smaller than if the books could be in any order? How is this any different from throwing the books away?

Activity 28

(c) You have $P(26, 5) = \binom{26}{5} \cdot 5!$. You know that $P(26, 5) = \frac{26!}{21!}$, so you should be able to easily "solve" for $\binom{26}{5}$.

Exercise 31 The mn looks like an application of the product principle. Say you had m bow ties and n fezzes. What question should you ask, and why does the other side also answer this? Alternatively, consider how this identity relates to that of Example 1.2.9.

Exercise 32 You will want to consider a lot of lattice paths, not all ending at the same point. In fact, what do all the lattice paths have in common?

Exercise 33 The answer looks like it might be almost the sum of a row in Pascal's triangle. Why does this make sense? You could think of including, or excluding numbers.

Exercise 34 Bit strings are nice here. What could the last bit of the bit string be?

Exercise 35 Try bit strings here. What length and weight are we looking at? Where might the first (left-most) one be?

Exercise 36 The identity you should try to prove is

$$\binom{2}{2}\binom{n}{2} + \binom{3}{2}\binom{n-1}{2} + \cdots + \binom{n}{2}\binom{2}{2} = \binom{n+3}{5}.$$

Exercise 37 In how many ways can you arrange the $n+1$ numbers $0, 1, 2, \ldots, n$ so that they are *not* in ascending order?

Exercise 38 Look back at Example 1.2.10. This one is easier.

Exercise 39 Think of partitioning the $n!$ permutations.

Exercise 41 Using Example 1.2.18 many times the left-hand side becomes can be decomposed as,

$$1^2 + 2^2 + 3^2 + \cdots + n^2 = \binom{n+1}{3} + \binom{n+2}{3}$$

$$1^2 + 2^2 + \cdots + (n-1)^2 = \binom{n}{3} + \binom{n+1}{3}$$

$$1^2 + 2^2 + \cdots + (n-2)^2 = \binom{n-1}{3} + \binom{n}{3}$$

$$\vdots$$

$$1^2 + 2^2 = \binom{3}{3} + \binom{4}{3}$$

$$1^2 = \binom{3}{3}.$$

What does this have to do with the problem and why does it help? Add all the columns!

Activity 43

(a) One rule of this type would be, two permutations are equivalent if they start with the same letter. This is not the one you want though.

(c) You might think about the usual way you would write a subset. Essentially what this question is asking is for you to pick a representative for each equivalence class.

Exercise 44 The problem suggests that you think about how to get a list from a seating arrangement. Could every list of n distinct people come from a seating chart? How many lists of n distinct people are there? How many lists could we get from a given seating chart by taking different starting places?

Additional Hint. For a different way of doing the problem, suppose that you have chosen one person, say the first one in a list of the people in alphabetical order by name. Now seat that person. Does it matter where they sit? In how many ways can you seat the remaining people? Does it matter where the second person in alphabetical order sits?

Exercise 45 How could we get a list of beads from a necklace?

Additional Hint. When we cut the necklace and string it out on a table, there are $2n$ lists of beads we could get. Why is it $2n$ rather than n?

Exercise 46

(a) You might first choose the pairs of people. You might also choose to make a list of all the people and then take them by twos from the list.

(b) You might first choose ordered pairs of people, and have the first person in each pair serve first. You might also choose to make a list of all the people and then take them by twos from the list in order.

Exercise 47 It might be helpful to just draw some pictures of the possible configurations. There aren't that many.

Exercise 48 A much easier question would be, how many anagrams of the word "anbgrcm"? Then use the quotient principle.

Activity 49

(a) This is not intended to be a trick question. In fact, this example would be a good one for thinking about how permutations and combinations are related in general.

(b) The point of this question is to push back against your conception of what "order matters" means. Since you know the answers must be $\binom{9}{5}$ or $P(9,5)$, you should be able to answer correctly by deciding which set is bigger.

(c) You might as well assume that $k \leq n$ (otherwise the answers would both be 0). If you are stuck, write out some examples of each using two-line notation for the functions. What decisions do you need to make?

Exercise 50 Think about what $P(n,k)$ would count when building a bit string. Why does it make sense to quotient out by $k!$?

Exercise 51 They key here is to think about the set of outcomes that we are counting. Ask yourself, the order of what?

Activity 52

(a) You should be able to use the multiplicative principle and nothing else here.

(b) Maybe the flavors are chocolate, vanilla, and strawberry. Some of the outcomes are ccv and csv, but notice that we would not also include cvc or vcs because in the blender, order doesn't matter.

Activity 57

(c) If you decide to put it on a shelf that already has a book, you have two choices of where to put it on that shelf.

(e) Among all the places you could put books, on all the shelves, how many are to the immediate left of some book? How many other places are there?

Activity 58

(b) Note, there are 5 exclamation marks, but 6 shelves.

(c) One such string is *!!***!*!*!, which looks a lot like 01100010101.

Activity 59

(a) The multiset $\{1, 2, 2, 3\}$ should correspond to the string *| * *|*.

Activity 62 How can you make sure that each shelf gets at least one book before you start the process described in Activity 57?

Exercise 63 We already know how to place k distinct books onto n distinct shelves so that each shelf gets at least one. Suppose we replace the distinct books with identical ones. If we permute the distinct books before replacement, does that affect the final outcome? There are other ways to solve this problem.

Exercise 66 Imagine taking a stack of k books, and breaking it up into stacks to put into the boxes in the same order they were originally stacked. If you are going to use n boxes, in how many places will you have to break the stack up into smaller stacks, and how many ways can you do this?

Additional Hint. How many different bookcase arrangements correspond to the same way of stacking k books into n boxes so that each box has at least one book?

Exercise 69 How many new handshakes occur when a new person enters the room?

Exercise 70 To determine a bijection $f : [n] \to [n]$, you could first say what $f(n)$ is. What do you still need to do? Your answer to this question should be one single step (that inductively, you know how to do).

Exercise 71 Suppose you already knew the number of moves needed to solve the puzzle with $n - 1$ rings.

Exercise 72 If we have $n - 1$ circles drawn in such a way that they define r_{n-1} regions, and we draw a new circle, each time it crosses another circle, except for the last time, it finishes dividing one region into two parts and starts dividing a new region into two parts.

Additional Hint. Compare r_n with the number of subsets of an n-element set.

Exercise 73

(a) List out the first few values of a_n by enumerating all paths.

Exercise 74 Your recurrence relation will not have a fixed number of terms, but rather be in terms of all previous terms.

Additional Hint. Fix an edge of the polygon. There will be exactly $n-2$ triangles this edge could be part of. For each of those, can you count the number of triangles that can be made of what is left?

Exercise 75 Try this out with some different initial conditions. Once you see what the recurrence should be, proving it by induction should be easy.

Activity 76 If you write out the sequence, it would be easy to guess at a closed formula, but the point here is to *use* the recurrence relation somehow. Consider rewriting the recurrence relation as $a_n - a_{n-1} = 2$. Note that $a_n - a_0 = (a_n - a_{n-1}) + (a_{n-1} + a_{n-2}) + \cdots + (a_2 - a_1) + (a_1 - a_0)$.

Activity 77 Again, if you write out the sequence, it would be easy to guess at a closed formula. To see how the recurrence relation is used, try substituting in earlier values. For example, $a_2 = 3a_1$, but $a_1 = 3a_0$, so $a_2 = 3(3a_0) = 3^2 a_0$.

Activity 79

(a) You will need to decide how many terms are on the right-hand side.

Exercise 82 Compute $S_n - 5S_n$

Exercise 84

(a) The number of plates that have 2 numerals and (therefore) 4 letters is $3^2 \cdot 4^4$, as there are 3 choices for each of the numerals, and 4 choices for each of the letters.

Activity 85

(a) You could do a full proof by induction here, but really you are just plugging these in to see if the recurrence works for them.

(c) If you do the substitution, you will get a degree n polynomial equation in the variable r. Solving that equation amounts to finding the roots of the polynomial.

Exercise 87 You will need to use the quadratic formula to find the roots, and some careful algebra to find the coefficients. The closed formula will include $\frac{1+\sqrt{5}}{2}$, the *Golden Ratio*.

Exercise 95 You should find that $Q^n = \begin{pmatrix} F_{n-1} & F_n \\ F_n & F_{n+1} \end{pmatrix}$. Now prove this by induction.

Exercise 99 Show first that $\frac{1}{F_{n-1}F_{n+1}} = \frac{1}{F_{n-1}F_n} - \frac{1}{F_n F_{n+1}}$.

Exercise 102 Make a table of Fibonacci numbers to get your conjecture.

Exercise 106 Use Q^{5n+5}.

Activity 110 Here we enumerate the number of workable sequences of n H's and n D's such that at each point in the sequence the number of H's is not less than the number of D's. Try writing out all HD sequences and see which ones are acceptable and which are not.

Activity 112 Try working recursively. For a product of $n+1$ symbols $a_1 a_2 \ldots a_{n+1}$, break it at the kth symbol:

$$(a_1 a_2 a_3 \ldots a_k)(a_{k+1} \ldots a_{n+1}).$$

Activity 113 You might have found a recursion for the sequence of answers in Exercise 74.

Activity 114 There should be 5 rooted binary trees with 4 leaves (the $n = 3$ case).

Activity 115

(a) If you put the numbers into the tableau in order, for each number, you must decide to put it in the top row or the bottom row. Do you have any other choices? What would constitute a mistake?

(b) One way to parenthesize the product $abcd$ is $a((bc)d)$, but this is really $(a((bc)d))$, so that the outer set set of parentheses belong with the product of a and $((bc)d)$. Further, not that if we wrote this just as $(a((bcd)$, there would be only one way to insert the right parentheses that would make the product parse correctly.

Exercise 116 Think about how students are taught to factor numbers.

Activity 117

(b) Look at rooted binary trees. What happens when you remove the root? You could also use the parenthesizing model by asking where "main product" occurs.

(c) Fix an edge of the polygon. There will be exactly $n - 2$ triangles this edge could be part of. For each of those, can you count the number of triangles that can be made of what?

Activity 118

(b) Given a path from $(0,0)$ to (n,n) which touches or crosses the line $y = x + 1$, how can you modify the part of the path from $(0,0)$ to the first touch of $y = x + 1$ so that the modified path starts instead at $(-1,1)$? The trick is to do this in a systematic way that will give you your bijection.

(c) A path either touches the line $y = x + 1$ or it doesn't. This partitions the set of paths into two blocks.

Exercise 119

(a) The algebraic formula for $\binom{n}{k}$ would be appropriate here.

Activity 133 Try drawing a Venn Diagram.

Exercise 138

(b) For each student, how big is the set of backpack distributions in which that student gets the correct backpack? It might be a good idea to first consider cases with $n = 3, 4,$ and 5.

Additional Hint. For each pair of students (say Mary and Jim, for example) how big is the set of backpack distributions in which the students in this pair get the correct backpack. What does the question have to do with unions or intersections of sets. Keep on increasing the number of students for which you ask this kind of question.

Exercise 139   Try induction.

Additional Hint.   We can apply the formula of Activity 133 to get

$$\left|\bigcup_{i=1}^{n} A_i\right| = \left|\left(\bigcup_{i=1}^{n-1} A_i\right) \cup A_n\right|$$

$$= \left|\bigcup_{i=1}^{n-1} A_i\right| + |A_n| - \left|\left(\bigcup_{i=1}^{n-1} A_i\right) \cap A_n\right|$$

$$= \left|\bigcup_{i=1}^{n-1} A_i\right| + |A_n| - \left|\bigcup_{i=1}^{n-1} A_i \cap A_n\right|$$

Exercise 140   Notice that it is straightforward to figure out how many ways we may pass out the apples so that child i gets five or more apples: give five apples to child i and then pass out the remaining apples however you choose. And if we want to figure out how many ways we may pass out the apples so that a given set C of children each get five or more apples, we give five to each child in C and then pass out the remaining $k - 5|C|$ apples however we choose.

Exercise 141   Start with two questions that can apply to any inclusion-exclusion problem. Do you think you would be better off trying to compute the size of a union of sets or the size of a complement of a union of sets? What kinds of sets (that are conceivably of use to you) is it easy to compute the size of? (The second question can be interpreted in different ways, and for each way of interpreting it, the answer may help you see something you can use in solving the problem.)

Additional Hint.   Suppose we have a set S of couples whom we want to seat side by side. We can think of lining up $|S|$ couples and $2n - 2|S|$ individual people in a circle. In how many ways can we arrange this many items in a circle?

Exercise 142   Reason somewhat as you did in Exercise 141, noting that if the set of couples who do sit side-by-side is nonempty, then the sex of the person at each place at the table is determined once we seat one couple in that set.

Additional Hint.   Think in terms of the sets A_i of arrangements of people in which couple i sits side-by-side. What does the union of the sets A_i have to do with the problem?

Activity 143

(a)   Note that 2341 should be circled, but 2314 should not be circled.

Exercise 146

(c)   The Taylor series for e^x is $1 + x + \frac{1}{2!}x^2 + \frac{1}{3!}x^3 + \cdots$. What happens if you replace x with -1?

Exercise 159   For example, to get the cost of the fruit selection APB you would want to get $x^{20} x^{25} x^{30} = x^{75}$.

Exercise 161   Consider the example with $n = 2$. Then we have two variables, x_1 and x_2. Forgetting about x_2, what sum says we either take x_1 or we don't?

Forgetting about x_1, what sum says we either take x_2 or we don't? Now what product says we either take x_1 or we don't *and* we either take x_2 or we don't?

Activity 163 For the last two questions, try multiplying out something simpler first, say $(a_0 + a_1 x + a_2 x^2)(b_0 + b_1 x + b_2 x^2)$. If this problem seems difficult the part that seems to cause students the most problems is converting the expression they get for a product like this into summation notation. If you are having this kind of problem, expand the product $(a_0 + a_1 x + a_2 x^2)(b_0 + b_1 x + b_2 x^2)$ and then figure out what the coefficient of x^2 is. Try to write that in summation notation.

Activity 164 Write down the formulas for the coefficients of x^0, x^1, x^2 and x^3 in

$$\left(\sum_{i=0}^{n} a_i x^i \right) \left(\sum_{j=0}^{m} b_j x^j \right).$$

Activity 165 How is this problem different from Activity 164? Is this an important difference from the point of view of the coefficient of x^k?

Activity 166

(b) You might try applying the product principle for generating functions to an appropriate power of the generating function you got in the first part of this problem.

Additional Hint. In Theorem 1.3.3 we have a formula for the number of k-element multisets chosen from an n-element set. Suppose you use this formula for a_k in $\sum_{k=0}^{\infty} a_k x^k$. What do you get the generating function for?

Activity 169 The only tricky thing here is that we have $\frac{1}{(1+x)^n}$ instead of $\frac{1}{(1-x)^n}$ which we know something about already. Note that $1 + x = 1 - (-x)$.

Additional Hint. Can you see a way to use Activity 168?

Activity 170

(a) You will get

$$\frac{1}{(1-x)^2} = 1 + 2x + 3x^2 + 4x^3 + \cdots.$$

Exercise 171 Look for a power of a polynomial to get started.

Additional Hint. The polynomial referred to in the first hint is a quotient of two polynomials. The power of the denominator can be written as a power series.

Exercise 172 You should have solved this problem more in general in Exercise 140.

Exercise 173 Interpret Exercise 171 in terms of multisets.

Exercise 176 You may run into a product of the form $\sum_{i=0}^{\infty} a^i x^i \sum_{j=0}^{\infty} b^j x^j$. Note that in the product, the coefficient of x^k is $\sum_{i=0}^{k} a^i b^{k-i} = \sum_{i=0}^{k} \frac{a^i}{b^i}$.

Exercise 179 Our recurrence becomes $a_n = a_{n-1} + a_{n-2}$, which of course is the Fibonacci recurrence.

Activity 182

$$\frac{5x+1}{(x-3)(x-5)} = \frac{cx+5c+dx-3d}{(x-3)(x-5)}$$

gives us

$$5x = cx + dx$$
$$1 = 5c - 3d.$$

Activity 183 To have

$$\frac{ax+b}{(x-r_1)(x-r_2)} = \frac{c}{x-r_1} + \frac{d}{x-r_2}$$

we must have

$$cx - r_2c + dx - r_1d = ax + b.$$

Exercise 185 $3 - 4x + x^2 = (1-x)(3-x)$. The power series for $\frac{A}{1-x}$ should be easy to express, but $\frac{B}{3-x}$ might make more sense written as $\frac{B/3}{1-\frac{1}{3}x}$.

Exercise 187 You can save yourself a tremendous amount of frustrating algebra if you arbitrarily choose one of the solutions and call it r_1 and call the other solution r_2 and solve the problem using these algebraic symbols in place of the actual roots.[1] Not only will you save yourself some work, but you will get a formula you could use in other problems. When you are done, substitute in the actual values of the solutions and simplify.

Exercise 188

(a) Once again it will save a lot of tedious algebra if you use the symbols r_1 and r_2 for the solutions as in Exercise 187 and substitute the actual values of the solutions once you have a formula for a_n in terms of r_1 and r_2.

(d) Think about how the binomial theorem might help you.

Exercise 190

(b) Does the right-hand side of the recurrence remind you of some products you have worked with?

(c)
$$\frac{1 \cdot 3 \cdot 5 \cdots (2i-3)}{i!} = \frac{(2i-2)!}{(i-1)!2^i i!}.$$

Activity 193

(d) What are the possible sizes of parts?

Activity 195 The number of partitions of $[k]$ into n parts in which k is not in a block relates to the number of partitions of $k-1$ into some number of blocks in a way that involves n. With this in mind, review how you proved Pascal's (recurrence) equation.

Exercise 196 To see how many broken permutations of a k element set into n parts do not have k is a part by itself, ask yourself how many broken permutations of $[7]$ result from adding 7 to the one of the two permutations in the broken permutation $\{14, 2356\}$.

Exercise 198 You can think of a function as assigning values to the blocks of its partition. If you permute the values assigned to the blocks, do you always change the function?

Exercise 199 When you add the number of functions mapping onto J over all possible subsets J of N, what is the set of functions whose size you are computing?

Activity 201

(a) As to why this is a reasonable question, think of distributing k items to n recipients. All four expressions count the number of ways to do this under different restrictions.

(b) Here it is helpful to think about what happens if you delete the entire block containing k rather than thinking about whether k is in a block by itself or not.

Exercise 203 What are the possible sizes of parts?

Activity 204

(b) Careful here. The answer is NOT $\binom{5}{2}\binom{3}{2}$. Do you see why?

(e) If you found this formula for $S(k, k-2)$ in part (e), then pretend your friend claims that $S(k, k-2) = \binom{k}{3} + \frac{1}{2}\binom{k}{2}\binom{k-2}{2}$ and prove why that is correct too.

Activity 205

(b) While you might not get this formula by conisdering type vectors, you should be shooting to prove $S(k, k-3) = \binom{k}{4} + 10\binom{k}{5} + 16\binom{k}{6}$

Activity 206 Suppose we make a list of the k items. We take the first a_1 elements to be the blocks of size 1. How many elements do we need to take to get a_2 blocks of size two? Which elements does it make sense to choose for this purpose?

Activity 209 What if the question asked about six sandwiches and two distinct bags? How does having identical bags change the answer?

Activity 211 What if the j_i's don't add to k?

Additional Hint. Think about listing the elements of the k-element set and labeling the first j_1 elements with label number 1.

Activity 212 The sum principle will help here.

Activity 213 How are the relevant j_i's in the multinomial coefficients you use here different from the j_i's in the previous problem.

Activity 214 Think about how binomial coefficients relate to expanding a power of a binomial and note that the binomial coefficient $\binom{n}{k}$ and the multinomial coefficient $\binom{n}{k,n-k}$ are the same.

Activity 215

(a) In what way is this problem like counting functions?

Exercise 217 You have already done this as well, although not with x.

Exercise 237 How many compositions are there of k into n parts? What is the maximum number of compositions that could correspond to a given partition of k into n parts?

Exercise 239 How can you start with a partition of k and make it into a new partition of $k+1$ that is guaranteed to have a part of size one, even if the original partition didn't?

Activity 242

(a) Note that you get partitions always of the same number (which?) but into a different number of parts (why?).

Exercise 244 Since the recurrence is a sum of smaller partition numbers, we are claiming that we can break the partitions of k into n parts into different groups, and each group will correspond to a different collection of smaller partitions. How can you describe these groups geometrically?

Activity 247 Draw a line through the top-left corner and bottom-right corner of the top-left box.

Activity 248 The largest part of a partition is the maximum number of boxes in a row of its Young diagram. What does the maximum number of boxes in a column tell us?

Activity 249 Draw all self conjugate partitions of integers less than or equal to 8. Draw all partitions of integers less than or equal to 8 into distinct odd parts (many of these will have just one part). Now try to see how to get from one set of drawings to the other in a consistent way.

Activity 250 Draw the partitions of six into even parts. Draw the partitions of six into parts used an even number of times. Look for a relationship between one set of diagrams and the other set of diagrams. If you have trouble, repeat the process using 8 or even 10 in place of 6.

Exercise 251 Draw a partition of ten into four parts. Assume each square has area one. Then draw a rectangle of area 40 enclosing your diagram that touches the top of your diagram, the left side of your diagram and the bottom of your diagram. How does this rectangle give you a partition of 30 into four parts?

Exercise 252

(c) Consider two cases, $m' > m$ and $m' = m$.

(d) Consider two cases, $n' > n$ and $n' = n$.

Exercise 253 Suppose we take two repetitions of this complementation process. What rows and columns do we remove from the diagram?

Additional Hint. To deal with an odd number of repetitions of the complementation process, think of it as an even number plus 1. Thus ask what kind of partition gives us the partition of one into one part after this complementation process.

Exercise 255 In the case $k = 4$ and $n = 2$, we have $m = 5$. In the case $k = 7$ and $n = 3$, we have $m = 10$.

Activity 256 What can you do to a Young diagram for a partition of k into n distinct parts to get a Young diagram of a partition of $k - n$ into some number of distinct parts?

Exercise 257 For any partition of k into parts λ_1, λ_2, etc. we can get a partition of k into odd parts by factoring the highest power of two that we can from each λ_i, writing $\lambda_i = \gamma_i \cdot 2^k_i$. Why is γ_i odd? Now partition k into 2^{k_1} parts of size γ_1, 2^{k_2} parts of size γ_2, etc. and you have a partition of k into odd parts.

Exercise 258 Suppose we have a partition of k into distinct parts. If the smallest part, say m, is smaller than the number of parts, we may add one to each of the m largest parts and delete the smallest part, and we have changed the parity of the number of parts, but we still have distinct parts. On the other hand, suppose the smallest part, again say m, is larger than or equal to the number of parts. Then we can subtract 1 from each part larger than m, and add a part equal to the number of parts larger than m. This changes the parity of the number of parts, but if the second smallest part is $m + 1$, the resulting partition does not have distinct parts. Thus this method does not work. Further, if it did always work, the case $k \ne \frac{3j^2+j}{2}$ would be covered also. However you can modify this method by comparing m not to the total number of parts, but to the number of rows at the top of the Young diagram that differ by exactly one from the row above. Even in this situation, there are certain slight additional assumptions you need to make, so this hint leaves you a lot of work to do. (It is reasonable to expect problems because of that exceptional case.) However, it should lead you in a useful direction.

Exercise 259 How does the number of compositions of k into n distinct parts compare to the number of compositions of k into n parts (not necessarily distinct)? What do compositions have to do with partitions?

Activity 260 This is a good place to apply the product principle for picture enumerators. Don't be afraid to use technology to finish the problem.

Activity 261

(a) The product principle for generating functions helps you break the generating function into a product of ten simpler ones.

(b) m was 10 in the previous part of this problem.

Activity 262 Think about conjugate partitions.

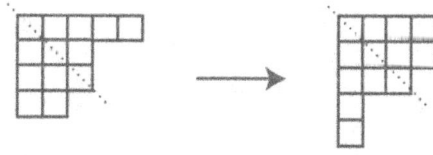

Exercise 263 Note that $1 + x^2 + x^4 + x^6 + \cdots$ in the generating function is really $1 + x^2 + x^{2 \cdot 2}$.

Exercise 266 In the power series $\sum_{j=0}^{\infty} x^{2ij}$, the $2ij$ has a different interpretation if you think of it as $(2i) \cdot j$ or if you think of it as $i \cdot (2j)$.

Exercise 267 Note that

$$\frac{1-x^2}{1-x} \cdot \frac{1-x^4}{1-x^2} \cdot \frac{1-x^6}{1-x^3} \cdot \frac{1-x^8}{1-x^4} = \frac{(1-x^6)(1-x^8)}{(1-x)(1-x^3)}.$$

Exercise 269 Note that $q^i + x^{3i} + x^{5i} + \cdots = x^i(1 + x^2 + x^4 + \cdots)$.

Activity 283

(a) Add something to your graph so you can "finish" the Euler path. Don't forget to consider the case that the start and end vertex are already adjacent.

(b) If there are no odd degree vertices, you are done. If there are two, you could connect them with an edge, except maybe they are already adjacent. What could you do then?

Activity 291

(a) Look at paths in the tree. Let $v_0, v_1, \ldots, v_n$ be a path of maximal length. What can you say about v_0 and v_n?

(b) What is the smallest number of vertices that make sense? You will need to prove that *all* trees with one vertex have the correct number of edges. This should be very easy.

(c) To get to a tree with $n - 1$ vertices, you need to *trim* T somehow. Which vertex should you get rid of? What will this do to the number of edges?

Activity 293 You can take $n = e = 1$ as your base case. For the inductive case, start with an arbitrary graph with n edges. Consider two cases: either G contains a cycle or it does not.

Activity 296

(a) If you assume P and conclude Q, but Q is true, can you say anything about P? What row(s) of the truth table for $P \to Q$ are you in?

Activity 298 The girth of any graph is at least 3.

Activity 299 Do a proof by contradiction. What is the smallest number of edges such a graph would possess?

Activity 310

(a) If you find a proper 5-coloring, could the chromatic number be more than 5? Could it be less than 5? What if you found a subgraph that you were sure had chromatic number 5?

Activity 311

(b) Modify your K_3-free graph with chromatic number 3.

(c) The smallest such graph contains 11 vertices. Try starting with C_5 and adding vertices creating a lot of odd cycles.

Activity 312 How many colors would you need to properly color the neighbors of v?

Activity 313 You could try a proof by induction, or more simply, just try coloring the graph in a *greedy* way.

Activity 318   Often when there is a counter-example, there is one with a good deal of symmetry. (Caution: there is a difference between often and always!) One way to help yourself get a symmetric example, if there is one, is to put 8 vertices into a circle. Then, perhaps, you might draw green edges in some sort of regular fashion until it is impossible to draw another green edge between any two of the vertices without creating a green triangle.

Activity 319   All that is left is to decide what can happen with K_9. First explain why no matter how the edges are colored red and blue, there is at least one vertex incident to an even number of red edges and an even number of blue edges. What could this even number of red edges be?

Activity 320   What you need to show is that if there are $R(m-1,n)+R(m,n-1)$ people in a room, then there are either m mutual acquaintances or n mutual strangers. As with earlier problems, it helps to start with a person and think about the number of people with whom this person is acquainted or nonacquainted. The generalized pigeonhole principle tells you something about these numbers.

Activity 321

(b)   If you could find four mutual acquaintances, you could assume person 1 is among them. And by the generalized pigeonhole principle and symmetry, so are two of the people to the first, second, fourth and eighth to the right. Now there are lots of possibilities for that fourth person. You now have the hard work of using symmetry and the definition of who is acquainted with whom to eliminate all possible combinations of four people. Then you have to think about non-acquaintances.

Activity 328

(c)   There is not much more to say now, except why b is not incident to any edge in M, and what the augmenting path would be.

Activity 329   Take A to be the 13 piles and B to be the 13 values. What would the matching condition need to say, and why is it satisfied.

APPENDIX D

SOLUTIONS TO SELECTED EXERCISES

Background Material

B.1.4.1. Solution.

(a) This is not a statement. It is an imperative sentence, but is not either true or false. It doesn't matter that this might actually be the rule or not. Note that "The rule is that all customers must wear shoes" *is* a statement.

(b) This is a statement, as it is either true or false. It is an atomic statement because it cannot be divided into smaller statements.

(c) This is again a statement, but this time it is molecular. In fact, it is a conjunction, as we can write it as "The customers wore shoes and the customers wore socks."

B.1.4.3. Solution.

(a) $P \wedge Q$.

(b) $P \to \neg Q$.

(c) Jack passed math or Jill passed math (or both).

(d) If Jack and Jill did not both pass math, then Jill did.

(e)   i. Nothing else.
      ii. Jack did not pass math either.

B.1.4.4. Solution.

(a) It is impossible to tell. The hypothesis of the implication is true. Thus the implication will be true if the conclusion is true (if 13 *is* my favorite number) and false otherwise.

(b) This is true, no matter whether 13 is my favorite number or not. Any implication with a true conclusion is true.

(c) This is true, again, no matter whether 13 is my favorite number or not. Any implication with a false hypothesis is true.

(d) For a disjunction to be true, we just need one or the other (or both) of the parts to be true. Thus this is a true statement.

(e) We cannot tell. The statement would be true if 13 is my favorite number, and false if not (since a conjunction needs both parts to be true to be true).

(f) This is definitely false. 13 is prime, so its negation (13 is not prime) is false. At least one part of the conjunction is false, so the whole statement is false.

(g) This is true. Either 13 is my favorite number or it is not, but whichever it is, at least one part of the disjunction is true, so the whole statement is true.

B.1.4.5. Solution. The main thing to realize is that we don't know the colors of these two shapes, but we do know that we are in one of three cases: We could have a blue square and green triangle. We could have a square that was not blue but a green triangle. Or we could have a square that was not blue and a triangle that was not green. The case in which the square is blue but the triangle is not green cannot occur, as that would make the statement false.

(a) This must be false. In fact, this is the negation of the original implication.

(b) This might be true or might be false.

(c) True. This is the contrapositive of the original statement, which is logically equivalent to it.

(d) We do not know. This is the converse of the original statement. In particular, if the square is not blue but the triangle is green, then the original statement is true but the converse is false.

(e) True. This is logically equivalent to the original statement.

B.1.4.6. Solution. The only way for an implication $P \to Q$ to be true but its converse to be false is for Q to be true and P to be false. Thus:

(a) False.

(b) True.

(c) False.

(d) True.

B.1.4.7. Solution. The converse is "If I will give you magic beans, then you will give me a cow." The contrapositive is "If I will not give you magic beans, then you will not give me a cow." All the other statements are neither the converse nor contrapositive.

B.1.4.9. Solution.

(a) Equivalent to the converse.

(b) Equivalent to the original theorem.

(c) Equivalent to the converse.

(d) Equivalent to the original theorem.

(e) Equivalent to the original theorem.

(f) Equivalent to the converse.

(g) Equivalent to the converse.

(h) Equivalent to the original theorem.

B.1.4.10. Solution.

(a) If you have lost weight, then you exercised.

(b) If you exercise, then you will lose weight.

(c) If you are American, then you are patriotic.

(d) If you are patriotic, then you are American.

(e) If a number is rational, then it is real.

(f) If a number is not even, then it is prime. (Or the contrapositive: if a number is not prime, then it is even.)

(g) If the Broncos don't win the Super Bowl, then they didn't play in the Super Bowl. Alternatively, if the Broncos play in the Super Bowl, then they will win the Super Bowl.

B.1.4.12. Solution. $P(5)$ is the statement "$3 \cdot 5 + 1$ is even", which is true. Thus the statement $\exists x P(x)$ is true (for example, 5 is such an x). However, we cannot tell anything about $\forall x P(x)$ since we do not know the truth value of $P(x)$ for *all* elements of the domain of discourse. In this case, $\forall x P(x)$ happens to be false (since $P(4)$ is false, for example).

B.1.4.14. Solution.

(a) The claim that $\forall x P(x)$ means that $P(n)$ is true no matter what n you consider in the domain of discourse. Thus the only way to prove that $\forall x P(x)$ is true is to check or otherwise argue that $P(n)$ is true for all n in the domain.

(b) To prove $\forall x P(x)$ is false all you need is one example of an element in the domain for which $P(n)$ is false. This is often called a counterexample.

(c) We are simply claiming that there is some element n in the domain of discourse for which $P(n)$ is true. If you can find one such element, you have verified the claim.

(d) Here we are claiming that no element we find will make $P(n)$ true. The only way to be sure of this is to verify that *every* element of the domain makes $P(n)$ false. Note that the level of proof needed for this statement is the same as to prove that $\forall x P(x)$ is true.

B.1.4.15. Suppose $P(x, y)$ is some binary predicate defined on a very small domain of discourse: just the integers 1, 2, 3, and 4. For each of the 16 pairs of these numbers, $P(x, y)$ is either true or false, according to the following table (x values are rows, y values are columns).

	1	2	3	4
1	T	F	F	F
2	F	T	T	F
3	T	T	T	T
4	F	F	F	F

For example, $P(1,3)$ is false, as indicated by the F in the first row, third column.

Solution.

(a) $\forall x \exists y P(x, y)$ is false because when $x = 4$, there is no y which makes $P(4, y)$ true.

(b) $\forall y \exists x P(x, y)$ is true. No matter what y is (i.e., no matter what column we are in) there is some x for which $P(x, y)$ is true. In fact, we can always take x to be 3.

(c) $\exists x \forall y P(x, y)$ is true. In particular $x = 3$ is such a number, so that no matter what y is, $P(x, y)$ is true.

(d) $\exists y \forall x P(x, y)$ is false. In fact, no matter what y (column) we look at, there is always some x (row) which makes $P(x, y)$ false.

B.1.4.16. Solution.

(a) $\neg \exists x (E(x) \wedge O(x))$.

(b) $\forall x (E(x) \rightarrow O(x + 1))$.

(c) $\exists x (P(x) \wedge E(x))$ (where $P(x)$ means "x is prime").

(d) $\forall x \forall y \exists z (x < z < y \vee y < z < x)$.

(e) $\forall x \neg \exists y (x < y < x + 1)$.

B.1.4.17. Solution.

(a) Any even number plus 2 is an even number.

(b) For any x there is a y such that $\sin(x) = y$. In other words, every number x is in the domain of sine.

(c) For every y there is an x such that $\sin(x) = y$. In other words, every number y is in the range of sine (which is false).

(d) For any numbers, if the cubes of two numbers are equal, then the numbers are equal.

B.2.5.1. Solution.

(a) 3.

(b) 3.

(c) 1.

(d) 1.

B.2.5.2. Solution.

(a) This is the set $\{3, 4, 5, \ldots\}$ since we need each element to be a natural number whose square is at least three more than 2. Since $3^2 - 3 = 6$ but $2^2 - 3 = 1$ we see that the first such natural number is 3.

(b) Here we don't specify that n must be a whole number at first, but the condition we place on it, that $n^2 - 5 \in \mathbb{N}$ does guarantee this (since square a non-integer always gives a non-integer, as does subtracting 5 from a non-integer). So in fact, we get the same set as we did in the previous part, and the smallest non-negative number for which $n^2 - 5$ is a natural numbers is 3.

Note that if we didn't specify $n \geq 0$ then any integer less than -3 would also be in the set, so there would not be a least element.

(c) This is the set $\{1, 2, 5, 10, \ldots\}$, namely the set of numbers that are the *result* of squaring and adding 1 to a natural number. ($0^2 + 1 = 1$, $1^2 + 1 = 2$, $2^2 + 1 = 5$ and so on.) Thus the least element of the set is 1.

(d) Now we are looking for natural numbers that are equal to taking some natural number, squaring it and adding 1. That is, $\{1, 2, 5, 10, \ldots\}$, the same set as the previous part. So again, the least element is 1.

B.2.5.3. Solution.

(a) 34. Note that $37 - 4 = 33$, but this calculation would not include 4 itself.

(b) 103. Again, you could compute this by $100 - (-2) + 1$, or simply count: 100 numbers from 1 through 100, plus -2, -1, and 0.

(c) 8. There are 8 primes not greater than 20: $\{2, 3, 5, 7, 11, 13, 17, 19\}$.

B.2.5.4. Solution. $\{2, 4\}$.

B.2.5.5. Solution. $\{1, 2, 3, 4, 5, 6, 8, 10\}$

B.2.5.6. Solution. 11.

B.2.5.7. Solution. There will be exactly 4 such sets: $\{2, 3, 4\}, \{1, 2, 3, 4\}, \{2, 3, 4, 5\}$ and $\{1, 2, 3, 4, 5\}$.

B.2.5.8. Solution.

(a) $A \cap B = \{3, 4, 5\}$.

(b) $A \cup B = \{1, 2, 3, 4, 5, 6, 7\}$.

(c) $A \setminus B = \{1, 2\}$.

(d) $A \cap \overline{(B \cup C)} = \{1\}$.

B.2.5.9. Solution.

(a) $A \cap B$ will be the set of natural numbers that are both at least 4 and less than 12, and even. That is, $A \cap B = \{x \in \mathbb{N} : 4 \leq x < 12 \wedge x \text{ is even}\} = \{4, 6, 8, 10\}$.

(b) $A \setminus B$ is the set of all elements that are in A but not B. So this is $\{x \in \mathbb{N} : 4 \leq x < 12 \wedge x \text{ is odd}\} = \{5, 7, 9, 11\}$.

Note this is the same set as $A \cap \overline{B}$.

B.2.5.11. Solution. For example, $A = \{2, 3, 5, 7, 8\}$ and $B = \{3, 5\}$.

B.2.5.12. Solution For example, $A = \{1, 2, 3\}$ and $B = \{1, 2, 3, 4, 5, \{1, 2, 3\}\}$

B.2.5.13. Solution.

(a) No.

(b) No.

(c) $2\mathbb{Z} \cap 3\mathbb{Z}$ is the set of all integers which are multiples of both 2 and 3 (so multiples of 6). Therefore $2\mathbb{Z} \cap 3\mathbb{Z} = \{x \in \mathbb{Z} : \exists y \in \mathbb{Z}(x = 6y)\}$.

(d) $2\mathbb{Z} \cup 3\mathbb{Z}$.

B.2.5.15. Solution.

(a) $A \cup \bar{B}$:

(b) $(\overline{A \cup B})$:

(c) $A \cap (B \cup C)$:

(d) $(A \cap B) \cup C$:

(e) $\bar{A} \cap B \cap \bar{C}$:

(f) $(A \cup B) \setminus C$:

B.2.5.17. Solution.

$$\mathcal{P}(A) = \{\emptyset, \{a\}, \{b\}, \{c\}, \{d\}, \{a,b\}, \{a,c\}, \{a,d\}, \{b,c\}, \{b,d\},$$
$$\{c,d\}\{a,b,c\}, \{a,b,d\}, \{a,c,d\}, \{b,c,d\}, \{a,b,c,d\}\}.$$

B.2.5.20. Solution.
For example, $A = \{1,2,3,4\}$ and $B = \{5,6,7,8,9\}$ gives $A \cup B = \{1,2,3,4,5,6,7,8,9\}$.

B.2.5.28. Solution.
We need to be a little careful here. If B contains 3 elements, then A contains just the number 3 (listed twice). So that would make $|A| = 1$, which would make $B = \{1,3\}$, which only has 2 elements. Thus $|B| \neq 3$. This means that $|A| = 2$, so B contains at least the elements 1 and 2. Since $|B| \neq 3$, we must have $|B| = 2$, which agrees with the definition of B.

Therefore it must be that $A = \{2,3\}$ and $B = \{1,2\}$

B.3.4.1. Solution.

(a) $f(1) = 4$, since 4 is the number below 1 in the two-line notation.

(b) Such an n is $n = 2$, since $f(2) = 1$. Note that 2 is above a 1 in the notation.

(c) $n = 4$ has this property. We say that 4 is a fixed point of f. Not all functions have such a point.

(d) Such an element is 2 (in fact, that is the only element in the codomain that is not in the range). In other words, 2 is not the image of any element under f; nothing is sent to 2.

B.3.4.2. Solution.

(a) This is neither injective nor surjective. It is not injective because more than one element from the domain has 3 as its image. It is not surjective because there are elements of the codomain (1, 2, 4, and 5) that are not images of anything from the domain.

(b) This is a bijection. Every element in the codomain is the image of *exactly* one element of the domain.

(c) This is a bijection. Note that we can write this function in two line notation as $f = \begin{pmatrix} 1 & 2 & 3 & 4 & 5 \\ 5 & 4 & 3 & 2 & 1 \end{pmatrix}$.

(d) In two line notation, this function is $f = \begin{pmatrix} 1 & 2 & 3 & 4 & 5 \\ 1 & 1 & 2 & 2 & 3 \end{pmatrix}$. From this we can quickly see it is neither injective (for example, 1 is the image of both 1 and 2) nor surjective (for example, 4 is not the image of anything).

B.3.4.5. Solution. There are 8 functions, including 6 surjective and zero injective funtions.

B.3.4.7. Solution.

(a) f is not injective, since $f(2) = f(5)$; two different inputs have the same output.

(b) f is surjective, since every element of the codomain is an element of the range.

(c) $f = \begin{pmatrix} 1 & 2 & 3 & 4 & 5 \\ 3 & 2 & 4 & 1 & 2 \end{pmatrix}$.

B.3.4.9. Solution. $f(10) = 1024$. To find $f(10)$, we need to know $f(9)$, for which we need $f(8)$, and so on. So build up from $f(0) = 1$. Then $f(1) = 2$, $f(2) = 4$, $f(3) = 8$, ... In fact, it looks like a closed formula for f is $f(n) = 2^n$. Later we will see how to prove this is correct.

B.3.4.10. Solution. For each case, you must use the recurrence to find $f(1)$, $f(2)$ ... $f(5)$. But notice each time you just add three to the previous. We do this 5 times.

(a) $f(5) = 15$.

(b) $f(5) = 16$.

(c) $f(5) = 17$.

(d) $f(5) = 115$.

B.3.4.12. Solution.

(a) f is injective, but not surjective (since 0, for example, is never an output).

(b) f is injective and surjective. Unlike in the previous question, every integers is an output (of the integer 4 less than it).

(c) f is injective, but not surjective (10 is not 8 less than a multiple of 5, for example).

(d) f is not injective, but is surjective. Every integer is an output (of twice itself, for example) but some integers are outputs of more than one input: $f(5) = 3 = f(6)$.

B.3.4.13. Solution.

(a) f is not injective. To prove this, we must simply find two different elements of the domain which map to the same element of the codomain. Since $f(\{1\}) = 1$ and $f(\{2\}) = 1$, we see that f is not injective.

(b) f is not surjective. The largest subset of A is A itself, and $|A| = 10$. So no natural number greater than 10 will ever be an output.

(c) $f^{-1}(1) = \{\{1\}, \{2\}, \{3\}, \ldots \{10\}\}$ (the set of all the singleton subsets of A).

(d) $f^{-1}(0) = \{\emptyset\}$. Note, it would be wrong to write $f^{-1}(0) = \emptyset$ - that would claim that there is no input which has 0 as an output.

(e) $f^{-1}(12) = \emptyset$, since there are no subsets of A with cardinality 12.

B.3.4.16. Solution.

(a) $|f^{-1}(3)| \leq 1$. In other words, either $f^{-1}(3)$ is the empty set or is a set containing exactly one element. Injective functions cannot have two elements from the domain both map to 3.

(b) $|f^{-1}(3)| \geq 1$. In other words, $f^{-1}(3)$ is a set containing at least one elements, possibly more. Surjective functions must have something map to 3.

(c) $|f^{-1}(3)| = 1$. There is exactly one element from X which gets mapped to 3, so $f^{-1}(3)$ is the set containing that one element.

B.3.4.17. Solution. X can really be any set, as long as $f(x) = 0$ or $f(x) = 1$ for every $x \in X$. For example, $X = \mathbb{N}$ and $f(n) = 0$ works.

B.3.4.21. Solution.

(a) f is injective.

Proof. Let x and y be elements of the domain $\mathbb{Z}$. Assume $f(x) = f(y)$. If x and y are both even, then $f(x) = x+1$ and $f(y) = y+1$. Since $f(x) = f(y)$, we have $x + 1 = y + 1$ which implies that $x = y$. Similarly, if x and y are both odd, then $x - 3 = y - 3$ so again $x = y$. The only other possibility is that x is even an y is odd (or visa-versa). But then $x + 1$ would be odd and $y - 3$ would be even, so it cannot be that $f(x) = f(y)$. Therefore if $f(x) = f(y)$ we then have $x = y$, which proves that f is injective. ∎

(b) f is surjective.

Proof. Let y be an element of the codomain $\mathbb{Z}$. We will show there is an element n of the domain ($\mathbb{Z}$) such that $f(n) = y$. There are two cases: First, if y is even, then let $n = y + 3$. Since y is even, n is odd, so $f(n) = n - 3 = y + 3 - 3 = y$ as desired. Second, if y is odd, then let $n = y - 1$. Since y is odd, n is even, so $f(n) = n + 1 = y - 1 + 1 = y$ as needed. Therefore f is surjective. ∎

B.3.4.22. Solution. Yes, this is a function, if you choose the domain and codomain correctly. The domain will be the set of students, and the codomain will be the set of possible grades. The function is almost certainly not injective, because it is likely that two students will get the same grade. The function might be surjective – it will be if there is at least one student who gets each grade.

B.3.4.24. Solution. This is not a function.

B.3.4.25. Solution. The recurrence relation is $f(n + 1) = f(n) + n$.

B.3.4.26. Solution. In general, $|A| \geq |f(A)|$, since you cannot get more outputs than you have inputs (each input goes to exactly one output), but you could have fewer outputs if the function is not injective. If the function is injective, then

$|A| = |f(A)|$, although you can have equality even if f is not injective (it must be injective *restricted* to A).

B.3.4.27. Solution. In general, there is no relationship between $|B|$ and $|f^{-1}(B)|$. This is because B might contain elements that are not in the range of f, so we might even have $f^{-1}(B) = \emptyset$. On the other hand, there might be lots of elements from the domain that all get sent to a few elements in B, making $f^{-1}(B)$ larger than B.

More specifically, if f is injective, then $|B| \geq |f^{-1}(B)|$ (since every element in B must come from at most one element from the domain). If f is surjective, then $|B| \leq |f^{-1}(B)|$ (since every element in B must come from at least one element of the domain). Thus if f is bijective then $|B| = |f^{-1}(B)|$.

B.4.5.1. Solution.

(a) P: it's your birthday; Q: there will be cake. $(P \vee Q) \to Q$

(b) Hint: you should get three T's and one F.

(c) Only that there will be cake.

(d) It's NOT your birthday!

(e) It's your birthday, but the cake is a lie.

B.4.5.2. Solution.

P	Q	$(P \vee Q) \to (P \wedge Q)$
T	T	T
T	F	F
F	T	F
F	F	T

B.4.5.3. Solution.

P	Q	$\neg P \wedge (Q \to P)$
T	T	F
T	F	F
F	T	F
F	F	T

If the statement is true, then both P and Q are false.

B.4.5.6. Solution. Make a truth table for each and compare. The statements are logically equivalent.

B.4.5.8. Solution.

(a) $P \wedge Q$.

(b) $(\neg P \vee \neg R) \to (Q \vee \neg R)$ or, replacing the implication with a disjunction first: $(P \wedge Q) \vee (Q \vee \neg R)$.

(c) $(P \wedge Q) \wedge (R \wedge \neg R)$. This is necessarily false, so it is also equivalent to $P \wedge \neg P$.

(d) Either Sam is a woman and Chris is a man, or Chris is a woman.

B.4.5.12. Solution. The deduction rule is valid. To see this, make a truth table which contains $P \vee Q$ and $\neg P$ (and P and Q of course). Look at the truth value of Q in each of the rows that have $P \vee Q$ and $\neg P$ true.

B.4.5.16. Solution.

(a) $\forall x \exists y (O(x) \wedge \neg E(y))$.

(b) $\exists x \forall y (x \geq y \vee \forall z (x \geq z \wedge y \geq z))$.

(c) There is a number n for which every other number is strictly greater than n.

(d) There is a number n which is not between any other two numbers.

B.5.6.1. Solution.

(a) For all integers a and b, if a or b is not even, then $a + b$ is not even.

(b) For all integers a and b, if a and b are even, then $a + b$ is even.

(c) There are numbers a and b such that $a + b$ is even but a and b are not both even.

(d) False. For example, $a = 3$ and $b = 5$. $a + b = 8$, but neither a nor b are even.

(e) False, since it is equivalent to the original statement.

(f) True. Let a and b be integers. Assume both are even. Then $a = 2k$ and $b = 2j$ for some integers k and j. But then $a + b = 2k + 2j = 2(k + j)$ which is even.

(g) True, since the statement is false.

B.5.6.2. Solution.

(a) Proof by contradiction. Start of proof: Assume, for the sake of contradiction, that there are integers x and y such that x is a prime greater than 5 and $x = 6y + 3$. End of proof: ... this is a contradiction, so there are no such integers.

(b) Direct proof. Start of proof: Let n be an integer. Assume n is a multiple of 3. End of proof: Therefore n can be written as the sum of consecutive integers.

(c) Proof by contrapositive. Start of proof: Let a and b be integers. Assume that a and b are even. End of proof: Therefore $a^2 + b^2$ is even.

B.5.6.3. Solution.

(a) Direct proof.

   Proof. Let n be an integer. Assume n is even. Then $n = 2k$ for some integer k. Thus $8n = 16k = 2(8k)$. Therefore $8n$ is even. ∎

(b) The converse is false. That is, there is an integer n such that $8n$ is even but n is odd. For example, consider $n = 3$. Then $8n = 24$ which is even but $n = 3$ is odd.

B.5.6.4. Solution.

(a) This is an example of the pigeonhole principle. We can prove it by contrapositive.

Proof. Suppose that each number only came up 6 or fewer times. So there are at most six 1's, six 2's, and so on. That's a total of 36 dice, so you must not have rolled all 40 dice. ∎

(b) We can have 9 dice without any four matching or any four being all different: three 1's, three 2's, three 3's. We will prove that whenever you roll 10 dice, you will always get four matching or all being different.

Proof. Suppose you roll 10 dice, but that there are NOT four matching rolls. This means at most, there are three of any given value. If we only had three different values, that would be only 9 dice, so there must be 4 different values, giving 4 dice that are all different. ∎

B.6.5.1. Solution.

(a) If we have a number of beans ending in a 5 and we double it, we will get a number of beans ending in a 0 (since $5 \cdot 2 = 10$). Then if we subtract 5, we will once again get a number of beans ending in a 5. Thus if on any day we have a number ending in a 5, the next day will also have a number ending in a 5.

(b) If you don't *start* with a number of beans ending in a 5 (on day 1), the above reasoning is still correct but not helpful. For example, if you start with a number ending in a 3, the next day you will have a number ending in a 1.

(c) Part (b) is the base case and part (a) is the inductive case. If on day 1 we have a number ending in a 5 (by part (b)), then on day 2 we will also have a number ending in a 5 (by part (a)). Then by part (a) again, we will have a number ending in a 5 on day 3. By part (a) again, this means we will have a number ending in a 5 on day 4

The proof by induction would say that on *every* day we will have a number ending in a 5, and this works because we can start with the base case, then use the inductive case over and over until we get up to our desired n.

B.6.5.2. Solution.

Proof. We must prove that $1 + 2 + 2^2 + 2^3 + \cdots + 2^n = 2^{n+1} - 1$ for all $n \in \mathbb{N}$. Thus let $P(n)$ be the statement $1 + 2 + 2^2 + \cdots + 2^n = 2^{n+1} - 1$. We will prove that $P(n)$ is true for all $n \in \mathbb{N}$. First we establish the base case, $P(0)$, which claims that $1 = 2^{0+1} - 1$. Since $2^1 - 1 = 2 - 1 = 1$, we see that $P(0)$ is true. Now for the inductive case. Assume that $P(k)$ is true for an arbitrary $k \in \mathbb{N}$. That is, $1 + 2 + 2^2 + \cdots + 2^k = 2^{k+1} - 1$. We must show that $P(k + 1)$ is true (i.e., that $1 + 2 + 2^2 + \cdots + 2^{k+1} = 2^{k+2} - 1$). To do this, we start with the left-hand side of $P(k + 1)$ and work to the right-hand side:

$$1 + 2 + 2^2 + \cdots + 2^k + 2^{k+1} = 2^{k+1} - 1 + 2^{k+1} \quad \text{by inductive hypothesis}$$
$$= 2 \cdot 2^{k+1} - 1$$
$$= 2^{k+2} - 1$$

Thus $P(k + 1)$ is true so by the principle of mathematical induction, $P(n)$ is true for all $n \in \mathbb{N}$. ■

B.6.5.3. Solution.

Proof. Let $P(n)$ be the statement "$7^n - 1$ is a multiple of 6." We will show $P(n)$ is true for all $n \in \mathbb{N}$. First we establish the base case, $P(0)$. Since $7^0 - 1 = 0$, and 0 is a multiple of 6, $P(0)$ is true. Now for the inductive case. Assume $P(k)$ holds for an arbitrary $k \in \mathbb{N}$. That is, $7^k - 1$ is a multiple of 6, or in other words, $7^k - 1 = 6j$ for some integer j. Now consider $7^{k+1} - 1$:

$$7^{k+1} - 1 = 7^{k+1} - 7 + 6 \quad \text{by cleverness: } -1 = -7 + 6$$
$$= 7(7^k - 1) + 6 \quad \text{factor out a 7 from the first two terms}$$
$$= 7(6j) + 6 \quad \text{by the inductive hypothesis}$$
$$= 6(7j + 1) \quad \text{factor out a 6}$$

Therefore $7^{k+1} - 1$ is a multiple of 6, or in other words, $P(k + 1)$ is true. Therefore by the principle of mathematical induction, $P(n)$ is true for all $n \in \mathbb{N}$. ■

B.6.5.4. Solution.

Proof. Let $P(n)$ be the statement $1 + 3 + 5 + \cdots + (2n - 1) = n^2$. We will prove that $P(n)$ is true for all $n \geq 1$. First the base case, $P(1)$. We have $1 = 1^2$ which is true, so $P(1)$ is established. Now the inductive case. Assume that $P(k)$ is true for some fixed arbitrary $k \geq 1$. That is, $1 + 3 + 5 + \cdots + (2k - 1) = k^2$. We will now prove that $P(k + 1)$ is also true (i.e., that $1 + 3 + 5 + \cdots + (2k + 1) = (k + 1)^2$). We start with the left-hand side of $P(k + 1)$ and work to the right-hand side:

$$1 + 3 + 5 + \cdots + (2k - 1) + (2k + 1) = k^2 + (2k + 1) \quad \text{by ind. hyp.}$$
$$= (k + 1)^2 \quad \text{by factoring}$$

Thus $P(k + 1)$ holds, so by the principle of mathematical induction, $P(n)$ is true for all $n \geq 1$. ■

B.6.5.5. Solution.

Proof. Let $P(n)$ be the statement $F_0 + F_2 + F_4 + \cdots + F_{2n} = F_{2n+1} - 1$. We will show that $P(n)$ is true for all $n \geq 0$. First the base case is easy because $F_0 = 0$ and $F_1 = 1$ so $F_0 = F_1 - 1$. Now consider the inductive case. Assume $P(k)$ is true, that is, assume $F_0 + F_2 + F_4 + \cdots + F_{2k} = F_{2k+1} - 1$. To establish $P(k+1)$ we work from left to right:

$$\begin{aligned} F_0 + F_2 + \cdots + F_{2k} + F_{2k+2} &= F_{2k+1} - 1 + F_{2k+2} && \text{by ind. hyp.} \\ &= F_{2k+1} + F_{2k+2} - 1 \\ &= F_{2k+3} - 1 && \text{by recursive def.} \end{aligned}$$

Therefore $F_0 + F_2 + F_4 + \cdots + F_{2k+2} = F_{2k+3} - 1$, which is to say $P(k+1)$ holds. Therefore by the principle of mathematical induction, $P(n)$ is true for all $n \geq 0$. ∎

B.6.5.6. Solution.
Proof. Let $P(n)$ be the statement $2^n < n!$. We will show $P(n)$ is true for all $n \geq 4$. First, we check the base case and see that yes, $2^4 < 4!$ (as $16 < 24$) so $P(4)$ is true. Now for the inductive case. Assume $P(k)$ is true for an arbitrary $k \geq 4$. That is, $2^k < k!$. Now consider $P(k+1)$: $2^{k+1} < (k+1)!$. To prove this, we start with the left side and work to the right side.

$$\begin{aligned} 2^{k+1} &= 2 \cdot 2^k \\ &< 2 \cdot k! && \text{by the inductive hypothesis} \\ &< (k+1) \cdot k! && \text{since } k+1 > 2 \\ &= (k+1)! \end{aligned}$$

Therefore $2^{k+1} < (k+1)!$ so we have established $P(k+1)$. Thus by the principle of mathematical induction $P(n)$ is true for all $n \geq 4$. ∎

B.6.5.12. Solution. The only problem is that we never established the base case. Of course, when $n = 0$, $0 + 3 \neq 0 + 7$.

B.6.5.13. Solution.
Proof. Let $P(n)$ be the statement that $n + 3 < n + 7$. We will prove that $P(n)$ is true for all $n \in \mathbb{N}$. First, note that the base case holds: $0 + 3 < 0 + 7$. Now assume for induction that $P(k)$ is true. That is, $k + 3 < k + 7$. We must show that $P(k+1)$ is true. Now since $k + 3 < k + 7$, add 1 to both sides. This gives $k + 3 + 1 < k + 7 + 1$. Regrouping $(k+1) + 3 < (k+1) + 7$. But this is simply $P(k+1)$. Thus by the principle of mathematical induction $P(n)$ is true for all $n \in \mathbb{N}$. ∎

B.6.5.14. Solution. The problem here is that while $P(0)$ is true, and while $P(k) \to P(k+1)$ for *some* values of k, there is at least one value of k (namely $k = 99$) when that implication fails. For a valid proof by induction, $P(k) \to P(k+1)$ must be true for all values of k greater than or equal to the base case.

B.6.5.16. Solution. We once again failed to establish the base case: when $n = 0$, $n^2 + n = 0$ which is even, not odd.

B.6.5.19. Solution. The proof will by by strong induction.

Proof. Let $P(n)$ be the statement "n is either a power of 2 or can be written as the sum of distinct powers of 2." We will show that $P(n)$ is true for all $n \geq 1$.

Base case: $1 = 2^0$ is a power of 2, so $P(1)$ is true.

Inductive case: Suppose $P(k)$ is true for all $k < n$. Now if n is a power of 2, we are done. If not, let 2^x be the largest power of 2 strictly less than n. Consider $n - 2^x$, which is a smaller number, in fact smaller than both n and 2^x. Thus $n - 2^x$ is either a power of 2 or can be written as the sum of distinct powers of 2, but none of them are going to be 2^x, so the together with 2^x we have written n as the sum of distinct powers of 2.

Therefore, by the principle of (strong) mathematical induction, $P(n)$ is true for all $n \geq 1$. ∎

B.6.5.25. Solution. The idea here is that if we take the logarithm of a^n, we can increase n by 1 if we multiply by another a (inside the logarithm). This results in adding 1 more $\log(a)$ to the total.

Proof. Let $P(n)$ be the statement $\log(a^n) = n \log(a)$. The base case, $P(2)$ is true, because $\log(a^2) = \log(a \cdot a) = \log(a) + \log(a) = 2\log(a)$, by the product rule for logarithms. Now assume, for induction, that $P(k)$ is true. That is, $\log(a^k) = k \log(a)$. Consider $\log(a^{k+1})$. We have

$$\log(a^{k+1}) = \log(a^k \cdot a) = \log(a^k) + \log(a) = k \log(a) + \log(a),$$

with the last equality due to the inductive hypothesis. But this simplifies to $(k+1) \log(a)$, establishing $P(k+1)$. Therefore by the principle of mathematical induction, $P(n)$ is true for all $n \geq 2$. ∎

APPENDIX E

GNU Free Documentation License

Version 1.3, 3 November 2008

Copyright © 2000, 2001, 2002, 2007, 2008 Free Software Foundation, Inc. `<http://www.fsf.org/>`

Everyone is permitted to copy and distribute verbatim copies of this license document, but changing it is not allowed.

0. PREAMBLE.

The purpose of this License is to make a manual, textbook, or other functional and useful document "free" in the sense of freedom: to assure everyone the effective freedom to copy and redistribute it, with or without modifying it, either commercially or noncommercially. Secondarily, this License preserves for the author and publisher a way to get credit for their work, while not being considered responsible for modifications made by others.

This License is a kind of "copyleft", which means that derivative works of the document must themselves be free in the same sense. It complements the GNU General Public License, which is a copyleft license designed for free software.

We have designed this License in order to use it for manuals for free software, because free software needs free documentation: a free program should come with manuals providing the same freedoms that the software does. But this License is not limited to software manuals; it can be used for any textual work, regardless of subject matter or whether it is published as a printed book. We recommend this License principally for works whose purpose is instruction or reference.

1. APPLICABILITY AND DEFINITIONS.

This License applies to any manual or other work, in any medium, that contains a notice placed by the copyright holder saying it can be distributed under the terms of this License. Such a notice grants a world-wide, royalty-free license, unlimited in duration, to use that work under the conditions stated herein. The "Document", below, refers to any such manual or work. Any member of the public is a licensee, and is addressed as "you". You accept the license if you copy, modify or distribute the work in a way requiring permission under copyright law.

A "Modified Version" of the Document means any work containing the Document or a portion of it, either copied verbatim, or with modifications and/or translated into another language.

A "Secondary Section" is a named appendix or a front-matter section of the Document that deals exclusively with the relationship of the publishers or authors of the Document to the Document's overall subject (or to related matters) and contains nothing that could fall directly within that overall subject. (Thus, if the Document is in part a textbook of mathematics, a Secondary Section may not explain any mathematics.) The relationship could be a matter of historical connection with the subject or with related matters, or of legal, commercial, philosophical, ethical or political position regarding them.

The "Invariant Sections" are certain Secondary Sections whose titles are designated, as being those of Invariant Sections, in the notice that says that the Document is released under this License. If a section does not fit the above definition of Secondary then it is not allowed to be designated as Invariant. The Document may contain zero Invariant Sections. If the Document does not identify any Invariant Sections then there are none.

The "Cover Texts" are certain short passages of text that are listed, as Front-Cover Texts or Back-Cover Texts, in the notice that says that the Document is released under this License. A Front-Cover Text may be at most 5 words, and a Back-Cover Text may be at most 25 words.

A "Transparent" copy of the Document means a machine-readable copy, represented in a format whose specification is available to the general public, that is suitable for revising the document straightforwardly with generic text editors or (for images composed of pixels) generic paint programs or (for drawings) some widely available drawing editor, and that is suitable for input to text formatters or for automatic translation to a variety of formats suitable for input to text formatters. A copy made in an otherwise Transparent file format whose markup, or absence of markup, has been arranged to thwart or discourage subsequent modification by readers is not Transparent. An image format is not Transparent if used for any substantial amount of text. A copy that is not "Transparent" is called "Opaque".

Examples of suitable formats for Transparent copies include plain ASCII without markup, Texinfo input format, LaTeX input format, SGML or XML using a publicly available DTD, and standard-conforming simple HTML, PostScript or PDF designed for human modification. Examples of transparent image formats include PNG, XCF and JPG. Opaque formats include proprietary formats that can be read and edited only by proprietary word processors, SGML or XML for which the DTD and/or processing tools are not generally available, and the machine-generated HTML, PostScript or PDF produced by some word processors for output purposes only.

The "Title Page" means, for a printed book, the title page itself, plus such following pages as are needed to hold, legibly, the material this License requires to appear in the title page. For works in formats which do not have any title page as such, "Title Page" means the text near the most prominent appearance of the work's title, preceding the beginning of the body of the text.

The "publisher" means any person or entity that distributes copies of the Document to the public.

A section "Entitled XYZ" means a named subunit of the Document whose title either is precisely XYZ or contains XYZ in parentheses following text that translates XYZ in another language. (Here XYZ stands for a specific section name mentioned below, such as "Acknowledgements", "Dedications", "Endorsements", or "History".) To "Preserve the Title" of such a section when you modify the Document means that it remains a section "Entitled XYZ" according to this definition.

The Document may include Warranty Disclaimers next to the notice which states that this License applies to the Document. These Warranty Disclaimers are considered to be included by reference in this License, but only as regards disclaiming warranties: any other implication that these Warranty Disclaimers may have is void and has no effect on the meaning of this License.

2. VERBATIM COPYING.

You may copy and distribute the Document in any medium, either commercially or noncommercially, provided that this License, the copyright notices, and the license notice saying this License applies to the Document are reproduced in all copies, and that you add no other conditions whatsoever to those of this License. You may not use technical measures to obstruct or control the reading or further copying of the copies you make or distribute. However, you may accept compensation in exchange for copies. If you distribute a large enough number of copies you must also follow the conditions in section 3.

You may also lend copies, under the same conditions stated above, and you may publicly display copies.

3. COPYING IN QUANTITY.

If you publish printed copies (or copies in media that commonly have printed covers) of the Document, numbering more than 100, and the Document's license notice requires Cover Texts, you must enclose the copies in covers that carry, clearly and legibly, all these Cover Texts: Front-Cover Texts on the front cover, and Back-Cover Texts on the back cover. Both covers must also clearly and legibly identify you as the publisher of these copies. The front cover must present the full title with all words of the title equally prominent and visible. You may add other material on the covers in addition. Copying with changes limited to the covers, as long as they preserve the title of the Document and satisfy these conditions, can be treated as verbatim copying in other respects.

If the required texts for either cover are too voluminous to fit legibly, you should put the first ones listed (as many as fit reasonably) on the actual cover, and continue the rest onto adjacent pages.

If you publish or distribute Opaque copies of the Document numbering more than 100, you must either include a machine-readable Transparent copy along with each Opaque copy, or state in or with each Opaque copy a computer-network location from which the general network-using public has access to download using public-standard network protocols a complete Transparent copy of the Document, free of added material. If you use the latter option, you must take reasonably prudent steps, when you begin distribution of Opaque copies in quantity, to ensure that this Transparent copy will remain thus accessible at the stated location until at least one year after the last time you distribute an Opaque copy (directly or through your agents or retailers) of that edition to the public.

It is requested, but not required, that you contact the authors of the Document well before redistributing any large number of copies, to give them a chance to provide you with an updated version of the Document.

4. MODIFICATIONS.

You may copy and distribute a Modified Version of the Document under the conditions of sections 2 and 3 above, provided that you release the Modified Version under precisely this License, with the

Modified Version filling the role of the Document, thus licensing distribution and modification of the Modified Version to whoever possesses a copy of it. In addition, you must do these things in the Modified Version:

A. Use in the Title Page (and on the covers, if any) a title distinct from that of the Document, and from those of previous versions (which should, if there were any, be listed in the History section of the Document). You may use the same title as a previous version if the original publisher of that version gives permission.

B. List on the Title Page, as authors, one or more persons or entities responsible for authorship of the modifications in the Modified Version, together with at least five of the principal authors of the Document (all of its principal authors, if it has fewer than five), unless they release you from this requirement.

C. State on the Title page the name of the publisher of the Modified Version, as the publisher.

D. Preserve all the copyright notices of the Document.

E. Add an appropriate copyright notice for your modifications adjacent to the other copyright notices.

F. Include, immediately after the copyright notices, a license notice giving the public permission to use the Modified Version under the terms of this License, in the form shown in the Addendum below.

G. Preserve in that license notice the full lists of Invariant Sections and required Cover Texts given in the Document's license notice.

H. Include an unaltered copy of this License.

I. Preserve the section Entitled "History", Preserve its Title, and add to it an item stating at least the title, year, new authors, and publisher of the Modified Version as given on the Title Page. If there is no section Entitled "History" in the Document, create one stating the title, year, authors, and publisher of the Document as given on its Title Page, then add an item describing the Modified Version as stated in the previous sentence.

J. Preserve the network location, if any, given in the Document for public access to a Transparent copy of the Document, and likewise the network locations given in the Document for previous versions it was based on. These may be placed in the "History" section. You may omit a network location for a work that was published at least four years before the Document itself, or if the original publisher of the version it refers to gives permission.

K. For any section Entitled "Acknowledgements" or "Dedications", Preserve the Title of the section, and preserve in the section all the substance and tone of each of the contributor acknowledgements and/or dedications given therein.

L. Preserve all the Invariant Sections of the Document, unaltered in their text and in their titles. Section numbers or the equivalent are not considered part of the section titles.

M. Delete any section Entitled "Endorsements". Such a section may not be included in the Modified Version.

N. Do not retitle any existing section to be Entitled "Endorsements" or to conflict in title with any Invariant Section.

O. Preserve any Warranty Disclaimers.

If the Modified Version includes new front-matter sections or appendices that qualify as Secondary Sections and contain no material copied from the Document, you may at your option designate some or all of these sections as invariant. To do this, add their titles to the list of Invariant Sections in the Modified Version's license notice. These titles must be distinct from any other section titles.

You may add a section Entitled "Endorsements", provided it contains nothing but endorsements of your Modified Version by various parties — for example, statements of peer review or that the text has been approved by an organization as the authoritative definition of a standard.

You may add a passage of up to five words as a Front-Cover Text, and a passage of up to 25 words as a Back-Cover Text, to the end of the list of Cover Texts in the Modified Version. Only one passage of Front-Cover Text and one of Back-Cover Text may be added by (or through arrangements made by) any one entity. If the Document already includes a cover text for the same cover, previously added by you or by arrangement made by the same entity you are acting on behalf of, you may not add another; but you may replace the old one, on explicit permission from the previous publisher that added the old one.

The author(s) and publisher(s) of the Document do not by this License give permission to use their names for publicity for or to assert or imply endorsement of any Modified Version.

5. COMBINING DOCUMENTS.

You may combine the Document with other documents released under this License, under the terms defined in section 4 above for modified versions, provided that you include in the combination all of the Invariant Sections of all of the original documents, unmodified, and list them all as Invariant Sections of your combined work in its license notice, and that you preserve all their Warranty Disclaimers.

The combined work need only contain one copy of this License, and multiple identical Invariant Sections may be replaced with a single copy. If there are multiple Invariant Sections with the same name but different contents, make the title of each such section unique by adding at the end of it, in parentheses, the name of the original author or publisher of that section if known, or else a unique number. Make the same adjustment to the section titles in the list of Invariant Sections in the license notice of the combined work.

In the combination, you must combine any sections Entitled "History" in the various original documents, forming one section Entitled "History"; likewise combine any sections Entitled "Acknowledgements", and any sections Entitled "Dedications". You must delete all sections Entitled "Endorsements".

6. COLLECTIONS OF DOCUMENTS.

You may make a collection consisting of the Document and other documents released under this License, and replace the individual copies of this License in the various documents with a single copy that is included in the collection, provided that you follow the rules of this License for verbatim copying of each of the documents in all other respects.

You may extract a single document from such a collection, and distribute it individually under this License, provided you insert a copy of this License into the extracted document, and follow this License in all other respects regarding verbatim copying of that document.

7. AGGREGATION WITH INDEPENDENT WORKS.

A compilation of the Document or its derivatives with other separate and independent documents or works, in or on a volume of a storage or distribution medium, is called an "aggregate" if the copyright resulting from the compilation is not used to limit the legal rights of the compilation's users beyond what the individual works permit. When the Document is included in an aggregate, this License does not apply to the other works in the aggregate which are not themselves derivative works of the Document.

If the Cover Text requirement of section 3 is applicable to these copies of the Document, then if the Document is less than one half of the entire aggregate, the Document's Cover Texts may be placed on covers that bracket the Document within the aggregate, or the electronic equivalent of covers if the Document is in electronic form. Otherwise they must appear on printed covers that bracket the whole aggregate.

8. TRANSLATION.

Translation is considered a kind of modification, so you may distribute translations of the Document under the terms of section 4. Replacing Invariant Sections with translations requires special permission from their copyright holders, but you may include translations of some or all Invariant Sections in addition to the original versions of these Invariant Sections. You may include a translation of this License, and all the license notices in the Document, and any Warranty Disclaimers, provided that you also include the original English version of this License and the original versions of those notices and disclaimers. In case of a disagreement between the translation and the original version of this License or a notice or disclaimer, the original version will prevail.

If a section in the Document is Entitled "Acknowledgements", "Dedications", or "History", the requirement (section 4) to Preserve its Title (section 1) will typically require changing the actual title.

9. TERMINATION.

You may not copy, modify, sublicense, or distribute the Document except as expressly provided under this License. Any attempt otherwise to copy, modify, sublicense, or distribute it is void, and will automatically terminate your rights under this License.

However, if you cease all violation of this License, then your license from a particular copyright holder is reinstated (a) provisionally, unless and until the copyright holder explicitly and finally terminates your license, and (b) permanently, if the copyright holder fails to notify you of the violation by some reasonable means prior to 60 days after the cessation.

Moreover, your license from a particular copyright holder is reinstated permanently if the copyright holder notifies you of the violation by some reasonable means, this is the first time you have received notice of violation of this License (for any work) from that copyright holder, and you cure the violation prior to 30 days after your receipt of the notice.

Termination of your rights under this section does not terminate the licenses of parties who have received copies or rights from you under this License. If your rights have been terminated and not permanently reinstated, receipt of a copy of some or all of the same material does not give you any rights to use it.

10. FUTURE REVISIONS OF THIS LICENSE.

The Free Software Foundation may publish new, revised versions of the GNU Free Documentation License from time to time. Such new versions will be similar in spirit to the present version, but may differ in detail to address new problems or concerns. See http://www.gnu.org/copyleft/.

Each version of the License is given a distinguishing version number. If the Document specifies that a particular numbered version of this License "or any later version" applies to it, you have the option of following the terms and conditions either of that specified version or of any later version that has been published (not as a draft) by the Free Software Foundation. If the Document does not specify a version number of this License, you may choose any version ever published (not as a draft) by the Free Software Foundation. If the Document specifies that a proxy can decide which future versions of this License can be used, that proxy's public statement of acceptance of a version permanently authorizes you to choose that version for the Document.

11. RELICENSING.

"Massive Multiauthor Collaboration Site" (or "MMC Site") means any World Wide Web server that publishes copyrightable works and also provides prominent facilities for anybody to edit those works. A public wiki that anybody can edit is an example of such a server. A "Massive Multiauthor Collaboration" (or "MMC") contained in the site means any set of copyrightable works thus published on the MMC site.

"CC-BY-SA" means the Creative Commons Attribution-Share Alike 3.0 license published by Creative Commons Corporation, a not-for-profit corporation with a principal place of business in San Francisco, California, as well as future copyleft versions of that license published by that same organization.

"Incorporate" means to publish or republish a Document, in whole or in part, as part of another Document.

An MMC is "eligible for relicensing" if it is licensed under this License, and if all works that were first published under this License somewhere other than this MMC, and subsequently incorporated in whole or in part into the MMC, (1) had no cover texts or invariant sections, and (2) were thus incorporated prior to November 1, 2008.

The operator of an MMC Site may republish an MMC contained in the site under CC-BY-SA on the same site at any time before August 1, 2009, provided the MMC is eligible for relicensing.

APPENDIX F
LIST OF SYMBOLS

Symbol	Description	Page
K_n	the complete graph on n vertices	121
K_n	the complete graph on n vertices.	122
$K_{m,n}$	the complete bipartite graph of of m and n vertices.	122
C_n	the cycle on n vertices	122
P_n	the path on $n+1$ vertices	122
$\chi(G)$	the chromatic number of G	144
$\chi'(G)$	the chromatic index of G	149
$N(S)$	the set of neighbors of S.	153
$P, Q, R, S, \ldots$	propositional (sentential) variables	166
$\wedge$	logical "and" (conjunction)	166
$\vee$	logical "or" (disjunction)	166
$\neg$	logical negation	167
$\exists$	existential quantifier	174
$\forall$	universal quantifier	174
$\emptyset$	the empty set	184
$\mathcal{U}$	universal set (domain of discourse)	184
$\mathbb{N}$	the set of natural numbers	184
$\mathbb{Z}$	the set of integers	184
$\mathbb{Q}$	the set of rational numbers	184
$\mathbb{R}$	the set of real numbers	184
$\mathcal{P}(A)$	the power set of A	184
$\{,\}$	braces, to contain set elements.	184
$:$	"such that"	184
$\in$	"is an element of"	184
$\subseteq$	"is a subset of"	185
$\subset$	"is a proper subset of"	185
$\cap$	set intersection	185
$\cup$	set union	185
$\times$	Cartesian product	185
$\setminus$	set difference	185
$\bar{A}$	the complement of A	185
$|A|$	cardinality (size) of A	185
$A \times B$	the Cartesian product of A and B	189
$f(A)$	the image of A under f.	203
$f^{-1}(B)$	the inverse image of B under f.	203
$P(n)$	the nth case we are trying to prove by induction	237
42	the ultimate answer to life, etc.	238

Index

k-coloring, 143
modus ponens, 217

adjacent, 122

Bell Number, 90
biconditional, 167
bijection, 202
binomial coefficients, 7
bipartite, 122, 123
broken permutation, 39
Brooks' Theorem, 149

Canadians, 243
cardinality, 185, 187
Cartesian product, 189
cases, 232
Catalan number
    generating function for, 84
chromatic index, 149
chromatic number, 123, 144
circuit, 125
closed formula
    for a function, 198
codomain, 195
coefficient
    multinomial, 93
complement, 62, 188
complement of a partition, 106
complete graph, 121, 123
composition, 209
conditional, 167
conjugate of an integer partition, 104
conjunction, 166
connected, 121, 123
connectives, 166
    and, 166
    if and only if, 167
    implies, 167
    not, 167
    or, 166
contraction, 135
contradiction, 229
contrapositive, 170

    proof by, 228
converse, 170
convex, 140
counterexample, 231
cube, 139
cycle, 123, 125

De Morgan's laws, 215
deduction rule, 217
degree, 123
deletion, 135
derangement, 61, 63
derangement problem, 61
diagram
    of a partition
        Ferrers, 102
        Young, 102
difference, of sets, 188
direct proof, 227
disjunction, 166
Doctor Who, 182
domain, 195
double negation, 215

empty set, 184
Euler path, 123
existential quantifier, 174

faces, 131
Ferrers diagram, 102
Fibonacci numbers, 78, 83
Four Color Theorem, 144
free variable, 174
function, 195
    ordered, 34
    surjective
        and Stirling Numbers, 90
functions
    onto
        number of, 66

generating function, 71
geometric series, 74
girth, 138
graph, 122

Hall's Marriage Theorem, 154
Hamilton path, 129
hatcheck problem, 61

if and only if, 167
if..., then..., 167
implication, 167
inclusion and exclusion principle, 59
    for unions of sets, 62
inclusive or, 167
induced subgraph, 120
induction, 237, 239
inductive hypothesis, 239
injection, 201
integers, 183, 184
intersection, 185, 188
isomorphic, 117
isomorphism, 117
isomorphism class, 119

Lah number, 39
lattice path, 3
law of logic, 220
linear recurrence
    second order, 78
logical equivalence, 214
logically valid, *see* law of logic

matching, 152
matching condition, 153
menage problem, 63
monochromatic, 150
multigraph, 121, 123
multinomial coefficient, 93

natural numbers, 184
negation, 167
neighbors, 153
NP-complete, 130

one-to-one, 201
onto, 200
onto functions
    number of, 66
ordered function, 34

partial fractions
    method of, 81
partition of a set

type vector, 92
partition of an integer, 100
    conjugate of, 104
    decreasing list, 101
    Ferrers diagram, 102
    into n parts, 100
    self conjugate, 105
    type vector, 101
    Young diagram, 102
partitions of a set
    number of, 90
Pascal's triangle, 2
path, 123, 125
perfect graph, 148
permutation
    broken, 39
picture enumerator, 70
picture enumerators
    product principle for, 70
pigeonhole principle, 230
planar, 123, 131
Platonic solids, 140
polyhedron, 140
power set, 184, 186
predicate, 174
prime numbers, 226, 245
principle of inclusion and
    exclusion, 59
    for unions of sets, 62
product principle, 14
    picture enumerators, 70
proof by cases, 232
proof by contradiction, 229
proof by contrapositive, 228
proposition, 212

quantifiers
    exists, 174
    for all, 174

range, 195
rationals, 184
reals, 184
recurrence
    constant coefficient, 78
    linear, 78
    second order, 78
reference, self, *see* self reference

second order recurrence, 78

self reference, *see* reference, self
self-conjugate partition, 105
series
    geometric, 74
set, 182
set difference, 188
statement, 165
subgraph, 120, 123
    induced, 120
subset, 186
sum principle, 14, 59
surjection, 200
surjections
    number of, 66
surjective function
    counting, 90

tautology, 214

trail, 125
tree, 124
truth table, 212
truth value, 166
type vector for a partition of an
    integer, 101
type vector of a partition of a set, 92

union, 185, 188
universal quantifier, 174

Venn diagram, 190
vertex coloring, 124, 143
Vizing's Theorem, 149

walk, 124, 125

Young diagram, 102

Colophon

This book was authored in PreTeXt.

www.ingramcontent.com/pod-product-compliance
Lightning Source LLC
Chambersburg PA
CBHW081424220526
45466CB00008B/2261